REACTIVE SEPARATION PROCESSES

Edited by
Santi Kulprathipanja
UOP LLC, Research Center, Des Plaines, Illinois

CRC Press
Taylor & Francis Group
Boca Raton London New York

CRC Press is an imprint of the
Taylor & Francis Group, an informa business
A TAYLOR & FRANCIS BOOK

CRC Press
Taylor & Francis Group
6000 Broken Sound Parkway NW, Suite 300
Boca Raton, FL 33487-2742

© 2002 by Taylor & Francis Group, LLC
CRC Press is an imprint of Taylor & Francis Group, an Informa business

First issued in paperback 2019

No claim to original U.S. Government works

ISBN-13: 978-0-367-44716-8 (pbk)
ISBN-13: 978-1-56032-825-4 (hbk)

Visit the Taylor & Francis Web site at
http://www.taylorandfrancis.com

and the CRC Press Web site at
http://www.crcpress.com

REACTIVE
SEPARATION
PROCESSES

CONTENTS

PREFACE

Reactive separation processes are unique in that they combine the normally disjunct unit operations of reaction and separation into a single, simultaneous operation. The advantages of such technology are principally in energy and capital cost reductions, as well as in increased reaction efficiency. Additionally, reactive separation is sometimes the only method by which to effect a separation when conventional means such as distillation or extraction are not applicable.

Reactive separations are not new. The application of reactive separations dates back to the earliest days of gas handling, coal tar and petroleum refining, and chemicals production. Absorption and reaction have been practiced commercially for decades. However, R&D activities relating to other areas of reactive separation concepts were not fully explored until the late 1970s to the early 1980s. During this period, commercial applications of reactive distillation processes for methyl tertiary butyl ether (MTBE) and methyl acetate manufacture were introduced. Many other applications using catalytic distillation are currently commercialized. For example, in the oxygenated fuel additive market, this technology is available for the reaction of simple alcohols with light tertiary olefins.

In addition to the above-mentioned reactive distillation techniques, other general reactive separation processes include reactive extraction, reactive adsorption, reactive membrane separation, and reactive crystallization. These technologies have proven to be increasingly attractive to industry and academia alike, as evidenced by the rapid increase in the number of technical articles and patent publications in the last decade. Furthermore, as either co-chair or chair of the 1995 (Miami), 1996 (Chicago), 1997 (Los Angeles), 1998 (Miami), 1999 (Dallas), and 2000 (Los Angeles) AIChE Annual Meeting sessions entitled "Reactive Separation Processes," the editor has personally witnessed the increasing interest in academia and industry in this topic. I am pleased to have been invited by Taylor & Francis Publishers, Inc., to put this book together in response to that interest.

This book is intended for anyone with a scientific background and interest in the field of reactive separations. In addition, clear and simple descriptions combined with illustrated examples will also help those inexperienced with the subject matter to easily comprehend the technical information. The book should, therefore, serve as a useful reference for chemical engineers and chemists in the classroom, laboratory, and in the plant, as well as business managers familiar with reactions and separations.

The goal of this book is to document and summarize the available information in six known areas of reactive separation:

- reaction/distillation
- reaction/extraction
- reaction/absorption
- reaction/adsorption
- reaction/membrane
- reaction/crystallization

Its structure places emphasis primarily on applications, but fundamental principles and technical considerations in industry are also recognized for each technology. The individual reactive separation processes are illustrated using numerous documented research and development studies which describe their respective applications. It is my hope that this book will offer guidance in problem solving and serve to generate further inventive and novel ideas for industrial application.

Santi Kulprathipanja

ACKNOWLEDGMENTS

I am indebted to Lisa G. Ehmer, a previous Senior Acquisitions Editor of Taylor & Francis Publishers, Inc., for encouraging me to assemble this book. I would also like to thank Catherine M. Caputo for her assistance on the book guidelines.

My special thanks to Dr. Inara M. Brubaker and Ms. Marcia L. Poteete for their assistance in assembling this book. Inara's insight and commentary were of tremendous value in helping me review and compile the contents of this work. Marcia's patience, persistence, and publication skills were essential in formatting this book. The work would have been impossible without their help.

I would also like to thank my contributors for their efforts in assembling the material for this book: Professor Robert W. Carr, Dr. Hemant W. Dandekar, Stanley J. Frey, Dr. Jeffrey S. Kanel, Dr. Vaibhav V. Kelkar, Dr. William A. Leet, Professor Jerry H. Meldon, Professor Ka M. Ng, Dr. Ketan D. Samant, Professor José G. Sanchez Marcano, Dr. Gavin P. Towler, Professor Theodore T. Tsotsis, and Professor Vincent Van Brunt.

Finally, I wish to thank Dr. Robert H. Jensen for his encouragement, and Benjamin C. Spehlmann, Robert F. Anderson, John F. Spears, Jr., Ames Kulprathipanja, and Ann Kulprathipanja for their reviews of this book. I would also like to thank UOP LLC for their support and encouragement.

S. K.

CONTRIBUTORS

Robert W. Carr, Ph.D.
Department of Chemical
Engineering and Materials Science
151 Amundson Hall
University of Minnesota
421 Washington Avenue, S.E.
Minneapolis, MN 55455
USA

Hemant W. Dandekar, Ph.D.
UOP LLC
25 East Algonquin Road
Des Plaines, IL 60017-5016
USA

Stanley J. Frey
UOP LLC
25 East Algonquin Road
Des Plaines, IL 60017-5016
USA

Jeffrey S. Kanel, Ph.D.
Dow Chemical Company
Technical Center, Building 740
3200 Kanawha Turnpike
South Charleston, WV 25303
USA

Vaibhav V. Kelkar, Ph.D.
Department of Chemical
Engineering
University of Massachusetts
Amherst, MA 01003-3110
USA

Santi Kulprathipanja, Ph.D.
UOP LLC
25 East Algonquin Road
Des Plaines, IL 60017-5016
USA

William A. Leet, Ph.D.
UOP LLC
25 East Algonquin Road
Des Plaines, IL 60017-5016
USA

Jerry H. Meldon, Ph.D.
Department of Chemical and
Biological Engineering
Tufts University
4 Colby Street
Medford, MA 02155
USA

Ka M. Ng, Ph.D.
Department of Chemical
Engineering
University of Massachusetts
Amherst, MA 01003-3110
USA

Ketan D. Samant, Ph.D.
Department of Chemical
Engineering
University of Massachusetts
Amherst, MA 01003-3110
USA

José G. Sanchez Marcano, Ph.D.
Institut Européen des Membranes
IEMM, UMII, cc 047
Place Eugène Bataillon
34095 Montpellier Cedex 5
France

Gavin P. Towler, Ph.D.
UOP LLC
25 East Algonquin Road
Des Plaines, IL 60017-5016
USA

Theodore T. Tsotsis, Ph.D.
Department of Chemical
Engineering
University of Southern California
University Park
Los Angeles, CA 90089-1211
USA

Vincent Van Brunt, Ph.D.
Department of Chemical
Engineering
University of South Carolina
Swearingen Center
Columbia, SC 29208
USA

Chapter 1

REACTIVE SEPARATION PROCESSES

William A. Leet and Santi Kulprathipanja

1.1 Introduction

Reactive separation processes are unique in that they couple chemical reactions and physical separations into a single unit operation. This fusion of reaction and separation operations into one combined operation is prized for the simplicity and novelty this approach brings to the process flowsheet. These reactive separations are also coveted for the investment and operating cost savings garnered on successful scale-up to commercial operations. Reactive separation processes as a whole are not a new concept. Numerous applications have been commercialized for traditional separation methods over the course of six decades. However, the academic and industrial communities have taken renewed interest in the development and commercialization of reactive separations in recent years in response to economic opportunities and pressures exerted, which have caused the emergence of new industries and decline of existing industries, the emergence of new separation and process technologies, the demand for higher building-block purities for food, pharmaceutical, polymer, and electronics products, changes in the availability and pricing of key resources, and the growing concern for protecting our environment. This interest is reflected in the increased volume of literature issued on the development and design of reactive separations and the increasing number of technical conferences devoted to the subject.

The concept of the reactive separation process is schematically presented in Figure 1.1. In traditional process design (the sequential reaction/separation process in Figure 1.1), a chemical reactor is typically sequenced with a downstream separator. In this way, feed is first converted into valued products which are then isolated and recovered in the separator. To optimize the product yield and purity, the operating conditions of the reactor and the separator are varied to achieve optimal economic performance subject to prevailing constraints. In many cases, recycle streams are incorporated into the process to reprocess unreacted feed or intermediate products back through the reactor and separator to increase overall process yields. In contrast, the design of a reactive separation process (the integrated reaction/separation process in Figure 1.1) focuses on integrating the reactor and separator operations into a single process operation with simultaneous reaction and separation. In combining sequential processing steps into an integrated processing approach, additional processing benefits may be achieved. Quite often, these benefits include the elimination of one or more recycle streams which are associated with optimizing performance of the original sequential process configuration. More excitingly, the integration may lead to the design of a separation process which cannot be achieved with separate reactor and separator process flow elements.

The applications of reactive separation process design are numerous and span a broad range of process operations. The development and application of reactive separations is not new. For example, reactive absorption has long been practiced for the removal of acid gases in the petroleum production and refining industries, and acid/base extractions and reactive distillations have long been practiced in the purification of chemicals. However, mention the phrase "reactive separations" and chemists and engineers will likely

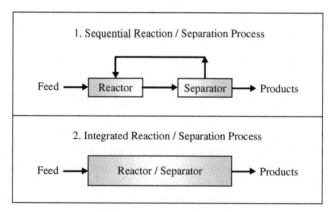

Figure 1.1:   Principle of combined reaction and separation system.

point to the more recent commercialization of reactive distillation technologies for the esterification of acetic acid to methyl acetate and the etherification of methanol with isobutylene to methyl tertiary butyl ether (MTBE) as examples of the art of reactive separation design.

These latter developments herald a renewal of interest in the development and commercialization of reactive separations. This trend is reflected in the rate at which the academic and industrial communities continue to expend effort to develop and apply reactive separations technology. A quick and rough measure of this trend is provided by scanning the number of citations issued each year on this subject in *Chemical Abstracts* using a keyword search (Table 1.1). For example, a quick citation search was conducted online for the subject of reactive distillation using keywords such as reactive distillation, distillation with reaction, simultaneous distillation with reaction, etc. This search yielded 263 publication citations and 123 patent citations from 1968 through the end of 1997. Breakdown of these listings by year in which the publications and patents were issued gives us a qualitative trend of the activity devoted to developments in this field as a function of time.

Figure 1.2 graphically presents the results of such a search for reactive separations in general for the 30-year time span from 1968 through 1997. The graph plots the cumulative number of citations at five-year intervals starting in 1972 for publications which specifically address reactive separations versus the year for which the citations were collected, on a semilog plot. The results are further segregated by type of process operations as follows: (1) reactive distillation, (2) extraction with reaction, (3) absorption with reaction, (4) adsorption with reaction, (5) reactive membrane separation, and (6) reactive crystallization. All six of these classes of reactive separations exhibit one to two orders of magnitude increase in the rate at which articles have been published in the literature over time.

This growth of interest in the subject of reactive separation processes is also reflected in the number of patents issued in each of the above areas. Using the same quick keyword search that was used to generate Figure 1.2, the growing interest can be roughly trended by examining the cumulative patent citation count for the 1968–1997 time frame. This data is plotted by class of separation in Figure 1.3 alongside the cumulative count for the number of article citations cited in Figure 1.2. The trends roughly match those observed for the open literature. The greatest number of patents was issued for reactive absorption technology development. The development of proprietary interests for the other five classes through the issue of patents lags the activity in the reactive absorption field by one to two orders of magnitude. Among these latter five classes, reactive membrane development followed by reactive distillation shows the greatest interest in terms of number of patents issued in these fields.

Table 1.1: Summary of Keywords Used to Conduct *Chemical Abstracts* Citation Search for Chapter 1 Figures.

Subject	Search Keywords	Search Results
Reactive distillation	Catalytic distillation or distillation, catalytic Catalysis with distillation or distillation with catalysis Reactive distillation or distillation, reactive Simultaneous (I)[a] reaction and (I) distillation	263 Publications from 1967–1998 123 Patents from 1967–1998
Extraction with reaction	Extraction with reaction or extraction with chemical reaction Reactive extraction Simultaneous (I) extraction and (I) reaction Combined (I) extraction and (I) reaction Extractive alkylation or extractive methylation Extraction, with reaction Catalysis (I) with extraction Reaction, with extraction	331 Publications from 1967–1998 32 Patents from 1967–1998
Absorption with reaction	Simultaneous absorption and reaction Gas treating or treatment or process Treating gas(es) or treatment gas(es) or processing gas(es) Absorptive reactive or absorptive reactivity Absorption reaction or absorption reactivity Absorption catalysis or catalytic or catalyst Absorptive catalysis or catalytic or catalyst Removal acid gas(es)	24,718 Publications from 1968–1998 22,425 Patents from 1968–1998

Table 1.1 (*continued*)

Subject	Search Keywords	Search Results
Adsorption with reaction	Adsorption reaction with or adsorption with reaction Reaction with adsorption or reaction, adsorption with Adsorption reaction Reaction, catalytic, adsorption (L)[b] (simultaneous or combined) Reactive adsorption Adsorption (L) reaction (L) simultaneous Simultaneous adsorption reaction Catalytic adsorption/obi (in selected sections of *Chemical Abstracts*)	688 Publications from 1968–1998 40 Patents from 1968–1998
Reactive membrane separation	Membrane reactor Reactive membrane Simultaneous reaction and membrane Membrane separation with reaction	1874 Publications from 1968–1998 262 Patents from 1968–1998
Reactive crystallization	Crystallization with reaction or reaction with crystallization Simultaneous crystallization (I) and reaction (I) Precipitation (I) reactive (but not in geological chemistry section of *Chemical Abstracts*) Crystallization (I) reactive and crystallization is an index heading	182 Publications from 1968–1998 22 Patents from 1968–1998

[a]I = search indexing term.
[b]L = search indexing terms linked.

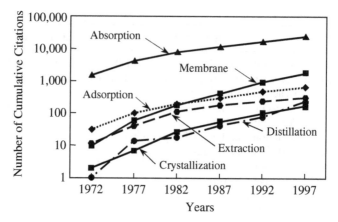

Figure 1.2: Number of *Chemical Abstracts* citations for publications on reactive separations from 1972 to 1997.

1.2   Advantages and Disadvantages

What is it that attracts the interest to develop these reactive separation processes? As difficult as it is to accurately trend the interest in the devel-

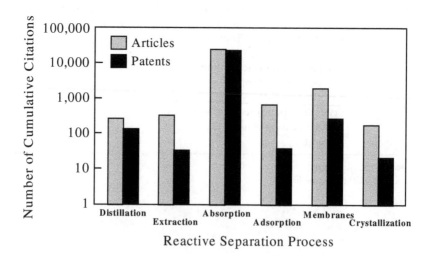

Figure 1.3: Number of *Chemical Abstracts* citations for articles vs. patents on reactive separations from 1972 to 1997.

opment of reactive separations, it is still more difficult to generalize the driving forces which lead to the development of these reactive separations. The simplest approach is to focus on the benefits and risks associated with successfully commercializing such technologies. A summary of the pros and cons associated with developing and operating reaction separation technologies is provided in Table 1.2. These pros and cons are further discussed below in roughly the same order.

1.2.1 Advantages

The potential advantages for implementation of reactive separations are numerous. Perhaps the most sought-after goals in the development of reactive separations are the linking of chemical "sources" and "sinks" to enhance reaction rates, conversions, or selectivities. Transport of desirable reaction products away from the catalyst may lead to enhanced reaction rates and increased feed conversion to products in equilibrium-controlled reaction regimes. Elimination of intermediates and suppression of side reactions may lead to higher yields of desirable byproducts, resulting in lower feedstock costs as a consequence of better feedstock utilization.

In some instances, the enhancement in reaction rate and selectivity in this way often allows the operator to reduce the reaction severity (e.g., reaction temperature, solution acidity, reaction pressure), to allow the process to be

Table 1.2: Advantages and Disadvantages of Reactive Separation Process Commercialization.

Advantages	Disadvantages
Enhanced reaction rates	Relatively new technology
Increased reaction conversion	Limited applications window
Enhanced reaction selectivity	Extensive experimental development
Reduced reaction severity	Complex modeling requirements
Increased catalyst life	Extensive equipment design effort
Simplified separations	Increased operational complexity
Improved product quality	Significant development costs
Heat integration benefits	Increased scale-up risks
Reduced equipment fouling/coking	
Inherently safer unit	
Reduced operating costs	
Reduced capital investment	
Novel process configurations	
Novel equipment designs	

operated under milder process conditions. Lowering the reaction severity may allow the operator to capture incremental improvements in reaction selectivity, product quality, catalyst life, run lengths, or even capital savings with downgrades in metallurgy to accommodate lower equipment design temperatures and pressures or less aggressive environments.

The coupling of mass transfer driving forces with reaction mechanisms often improves catalyst life. In many process applications, one of the primary catalyst deactivation mechanisms involves inhibition or "coking" of the catalyst by heavier, more polar, and generally undesirable reaction byproducts. The increased mass transfer driving forces realized in a reactive separator often lead to better catalyst irrigation and surface renewal with transport of potential catalyst inhibitors away from the catalyst surface. The enhanced selectivity achieved in many instances reduces the formation rate of the byproducts responsible for such catalyst inhibition. And in specific instances, the driving forces for separation may even inhibit the transport of the chemical precursors for such inhibitors to the catalyst surface in the first place. The combination of these effects often leads to better catalyst activity retention over time, leading to enhanced reactor productivity and lower catalyst costs.

The coupling of reaction and separation phenomena often leads to simplified separations. Conducting separations in zones where chemical reactions are taking place can lead to effectively higher local driving forces for separations leading to reductions in equipment size, elimination of recycle streams, and reductions in utility costs. Elimination of intermediates and suppression of side reactions may reduce purification requirements, leading to the implementation of simpler product purification systems. For example, suppression of byproduct formation may lead to suppression of undesirable azeotropes in reactive distillation. And in some cases, the coupling of reaction and separations may provide the means to conduct a separation where none is achievable by conventional means such as distillation or extraction.

The coupling of reaction and separation phenomena leads to coupled heat transfer effects. This leads to alternative design options for heat addition and removal in the design of reactive separation processes. In many cases, exothermic reactors (heat sources) can be coupled directly to energy-intensive separations like distillation columns (heat sinks) in the design of reactive separators. In such scenarios, heat energy can be transferred by direct contact as opposed to indirect transfer across heat exchange equipment between heat sources and sinks. In other cases, the coupling of heat transfer effects through reactive separator design opens new options for different and often times more efficient design options for heat integrating a process design.

Reduced equipment fouling and coking are often achieved in the design of reactive separators apart from that which affects catalyst life. Suppression of byproduct formation reactions often leads to reduced "polymer" or "heavies" formation in the reactor and subsequent deposition of higher molecular weight or less soluble byproducts on downstream processing equipment, with subsequent fouling of that equipment. Reduced fouling and coking of process heat exchangers result in higher apparent heat transfer coefficients. Reduced fouling and coking of distillation tower internals lead to sustained operation of the distillation tower at the design hydraulic and separation performance points. Such reductions in equipment fouling rates translate into longer run lengths between interruptions to operations to clean equipment to restore operating performance. This in turn leads to increased onstream efficiency, greater overall process throughput, and reduced turnaround, maintenance, and cleaning costs in operating the unit.

When the coupling of reactions and separations leads to a reduction in the number of pieces of process equipment required and substantial reduction in the sizing of such equipment, increased benefits may be achieved in terms of designing an inherently safer unit. If reactive separator design reduces the working inventory of catalyst or reactive or hazardous chemicals in the equipment, the unit is made "inherently safer" in terms of dealing with potential leaks, spills, and environmental releases. If the coupling of reactions and separations leads to suppression of byproduct reactions which are likely to exhibit runaway behavior, the reactive separator design will increase the inherent safety in the unit against severe process upsets.

The benefits above can also lead to reductions in capital investment through the consolidation of multiple pieces of process equipment into a single piece and/or through the elimination of process recycle streams. Impregnation of an adsorbent with a catalytic component for byproduct destruction and deposition on the adsorbent can lead to a simplified product recovery scheme in which a product recycle stream may be eliminated in favor of a slightly larger product treater vessel. This may result in the elimination of the process recycle and the equipment and extra upstream equipment capacity incorporated to handle the excess flow. Insertion of a packed catalyst bed with high hydraulic capacity into a distillation tower and injection of a reactant at trace levels can lead to conversion of a distillable reactor byproduct into a higher boiling component to break an azeotrope. Relocation of a reactor catalyst bed into a staged phase separation vessel can lead to elimination of the original reactor or separation vessel.

The benefits above can lead to significant operating cost reductions. Increased reactor productivity and product quality are reflected in improved feedstock utilization and lower feedstock costs per unit of product. Increased catalyst activity and life lead to reduced catalyst costs. Heat integration benefits result in lower utility costs. Reduced equipment fouling and coking lead to better onstream operation and equipment utilization with reduced process maintenance costs.

Still another advantage to reactive separations design is the "novelty" element involved. Creativity is required to develop the process, the catalyst, the chemical environment, and the process equipment to produce a viable reactive separator. The results are generally patentable in terms of process configuration and process equipment design. Such patents distinguish the owner's process from competitor's and allow the developer to secure the economic benefits of such inventions for internal application. The patents also open the door for the owner to profitably license the technology to others.

1.2.2   Disadvantages

The potential disadvantages for implementation of reactive distillations must also be considered. Reactive separations are generally regarded as "new" technologies. The development and commercialization of new technology is often avoided due to the perceived higher costs and higher risks associated with implementation.

The operating window associated with the development of reactive separations is generally more limited than development of separate reaction and separation systems in the conventional approach. The overlap in process conditions required to provide acceptable reaction and separations performance is often quite narrow, or in many cases, nonexistent. In many applications, the overlap is so small that application of the reactive separator concept is reduced to the role of product treating.

Development of reactive separation processes is usually a resource intensive effort. Applications of reactive separations are custom tailored to the problem to be addressed. Solutions that work for one application do not necessarily apply to the next similar situation. Roadmaps and recipes for developing such processes are not always available, although a great deal of effort has been devoted in recent years to developing methods for screening and for developing reactive separations for more general application. The collection of more and better experimental data is generally required to clarify and validate the principles involved and to establish the operating window for application of such a process.

The coupling of reaction and separations phenomena necessarily increases the complexity of the process-modeling effort with the coupling of the driving forces for reaction and separation. Successful modeling often requires closer examination of the relevant phenomena on a localized or micro scale and subsequent integration of the micromodel over the associated time and spatial framework to scale up the process. This is more involved from the viewpoint of model development, model discrimination, parameter fitting, and computational horsepower.

Extensive equipment design is often required to address the details of how the reaction and separation phenomena can be coupled into a single vessel. More care and development are required to address the performance issues of such equipment in the design and scale-up phase. Reactive separator design usually exhibits greater interaction among the design components in operation and restricted access to the components once the process unit is put into operation. The increased chances of failing to meet performance expectations on unit start-up and the reduced opportunity to access and fix any problems quickly after unit start-up are strong incentives to spend more effort on equipment design in the design phase. Quite often this requires the development of new reactor–separator internals to accomplish this.

The development of reactive separation processes generally leads to an increase in the complexity of process operations and the control systems instituted to regulate the process. The coupling of reactions and separations usually leads to a reduction in the number of control variables which can be used to manipulate the process. Often, traditional process control variables such as temperatures, flow rates, and pressures can no longer be used as control variables as they are embedded into the operation of the unit as "outcomes" rather than as "inputs." Thus, it becomes much more difficult to relate cause and effects through the control scheme. The complexity of the control network must often be increased to accommodate such indirect controls. New control variables must be identified, and in many cases, key process controls must be set up to infer or estimate process responses to tune process operations. Process control ranges are often restricted to a smaller window of operation. Process control system stability and robustness can be compromised such that small changes in design parameters in the operation of the unit may lead to drastic and unforeseen results if not properly modeled. The vulnerability of the system to upset may increase. Process unit start-up and shutdown procedures usually become more involved in an effort to preserve catalyst activity while initiating separation.

More importantly, the increased process design complexity and the accompanying demands on development resources to carry the development to satisfactory conclusion lead to higher initial development costs.

The associated risks of commercialization of new reactive separations technology are generally higher. The risks can be minimized with good upfront research and development. But the risks cannot be eliminated entirely. The risks of venturing into new territory remain. These risks are exaggerated by the loss of operating flexibility for the reactive separator relative to separate reactor and separator sections as well as the loss in equipment count or sizing with any equipment savings realized. Options for fixing performance shortfalls with changes in operating parameters or procedures are restricted by the loss in operating flexibility. By eliminating equipment in coupling the reactions and separations, there is less equipment with which to effect fixes on performance shortfalls. The combination of these two factors can lead to additional development costs in the lab and in the field to recover from performance shortfalls.

1.3 Applications Topics

This book is dedicated to the discussion of developments in the application of reactive separations technology to each of the six aforementioned fields of interest: (1) reactive distillation, (2) extraction with reaction, (3) absorption with reaction, (4) adsorption with reaction, (5) reactive membrane separation, and (6) reactive crystallization. In Chapters 2–7, each of these areas of applications is reviewed in more depth, with a review of the general history, design basics, and industrial applications associated with these applications.

Chapter 2, "Reactive Distillation," by Gavin P. Towler and Stanley J. Frey, describes how reactive distillation takes advantage of simultaneous reaction and multistage vapor–liquid separation. In reactive distillation, one or more components in the feedstock undergoes significant chemical reaction at distillation temperature and pressure to transform one or more components in the feedstock with simultaneous fractionation of the resulting components into two or more fractions. Reactive distillation for trace impurity removal has long been practiced through additive injection for selective complexation, hydrolysis, neutralization, and reduction reactions. Reactive distillation for the conversion and separation of major feed components dates back to the 1920s and 1930s with research into the application of reactive distillation for the production of esters (Backhaus, 1921; Keyes, 1932). Reactive distillation developments were subsequently applied to various processes including several nitration, sulfonation, and saponification reactions. But it is the more recent commercialization successes for the esterification of acetic acid to methyl acetate (Agreda et al., 1990) and the etherification of isobutylene to fuel-grade methyl *tert*-butyl ether (Lander et al., 1983) that have heralded an increase in the development and application of reactive distillation technology. The introduction of fixed beds of catalyst in the latter development has led to a surge of development activity in the extension of reactive distillation to other reversible and irreversible reactions (Terril et al., 1985; Kruel et al., 1999). Novel tower internal and heterogeneous catalyst developments continue to increase the process potential. The economic advantages for reactive distillation, compared to conventional processes, lie in enhanced reaction conversion and selectivity, improved product quality, heat integration, and reduced process complexity.

Chapter 3, "Extraction with Reaction," by Vincent Van Brunt and Jeffrey S. Kanel, focuses on liquid–liquid extraction—the unit operation which partitions at least one solute between two partially miscible liquid phases differing in bulk density by an order of magnitude or less. Extraction is an indirect separation process whose economics are usually not governed by the chemical extraction step alone. Therefore, extraction flowsheets showing the integrated separation schemes which purify the solute and regenerate solvent are needed. To enhance separation efficiency, extraction is often coupled with reaction. Alternatively, extraction may be used to influence the extent of an equilibrium-limited reaction by substantially changing the relative amount of a reactant or product. The second liquid phase can thus serve as a source of a reactant or a sink for a reaction product. Commercially, both reversible and irreversible reactions can be used with extraction. Reactions confined to a single liquid phase, taking place at the liquid–liquid interface, or occurring in both liquid phases are all candidates for reactive extraction. Furthermore, the reactions may promote the formation or dissolution of solid phases. Extraction with reaction is commonly practiced in industry. Enhancement of the process through application of acid–base chemistry has long been practiced in the use of acid and base extractions for the purification of chemicals and fuels. Enhancement of the extraction process through application of complexation chemistry on the extraction process has also been practiced for some time. For example, extraction with reaction has become a well-accepted means in hydrometallurgy following the introduction of selective chelating agents in the 1960s (Rydberg et al., 1992). The enhancement of the extraction process with stronger reactions has also been practiced for some time. Familiar examples include the commercialization of sulfuric acid alkylation, HF alkylation, and various oxidative sweetening processes in the petroleum refining industry (Bland and Davidson, 1967). Additional developments can be found for a host of applications in the petroleum, chemical, pharmaceuticals, environmental, metallurgical, and nuclear industries. Drivers for developments in this area include yield and product purity enhancements and a stronger emphasis on protection of the environment.

Chapter 4, "Absorption with Reaction," by Jerry H. Meldon, focuses on the use of liquid-phase chemical reaction to enhance the rate of and capacity for absorption of one or more components of a gaseous mixture. Absorption with reaction is the oldest of the six reactive separation concepts, dating back to the early days of gas treating. Industry has seen numerous applications of reactive absorption in more recent times in the gas production, petroleum refining, chemical, and nuclear industries. This pattern follows industrial interests in the drive to enact enhanced pollution control measures to reduce plant emissions to meet growing global concern over pollution. The theoretical challenge in developing reactive absorbers is to accurately model simultaneous mass transfer and chemical reaction (Danckwerts, 1970; Astarita et al., 1983; Zarzycki and Chacuk, 1993). Successful modeling facilitates reliable prediction of overall transfer rates, as well as selectivity in gas separations. Particularly challenging is design for selective separation of one gaseous species when a solvent may react with a second more concentrated species. The authors review current models of mass and heat transfer and reaction in absorbers and strippers. In addition, the authors review more practical issues involving absorber type and configuration selection criteria while emphasizing the impact of process engineering decisions on selectivity and economics.

Chapter 5, "Adsorption with Reaction," by Robert W. Carr and Hemant W. Dandekar, covers an area of much recent interest. Reactive adsorption is characterized by simultaneous chemical reaction and separation. In a single reactor–separator vessel, it is possible to obtain high-purity product directly from the reactor, with downstream purification requirements greatly reduced or even eliminated. This can result in reduced design complexity and capital cost. Developments in this area date back to the applications of simple chemical treatments such as the use of acidic clays for the removal of olefins in hydrocarbon streams by acid-catalyzed polymerization over the clay beds, and the use of solid KOH to remove sulfur from various hydrocarbon streams (Bland and Davidson, 1967). More recent developments have followed on the heels of discovery and development of better-engineered synthetic adsorbents and catalysts for applications in the food, biotechnology, pharmaceuticals, chemicals, refining, environmental, and nuclear industries. For example, molecular sieves and ion exchange resins, exchanged with silver ions, have been used to treat vapor and liquid streams to remove trace iodides by reaction and precipitation to facilitate safe operation of nuclear reactors (Pence and Macek, 1970), and to purify acetic acid produced by methanol carbonylation (Hilton, 1986; Kulprathipanja et al., 1999). Similar examples can be found in the development of impregnated activated carbon adsorbents (Tien, 1994) and adsorbents modified to anchor biologically based catalysts (e.g., biofilms and enzymes). Many reactive adsorption processes developed to date utilize traditional fixed-bed and fluidized/moving-bed adsorber designs. However, much recent development effort has been focused on the development of improved purification processes in the form of reactive chromatographic methods and simulated moving-bed technologies. This chapter is focused on leading-edge developments in the preparative chromatography and simulated moving-bed (SMB) technologies.

Chapter 6, "Reactive Membrane Separation," by José Sanchez Marcano and Theodore T. Tsotsis, focuses on this currently widely utilized technology. Membrane separations are attractive in catalytic reaction applications where the integrated separation step results in enhanced selectivity and/or yield. Applications include the petrochemical, food, pharmaceutical, biotechnology, chemical, and recently emerging fuel cell industries (Scott and Hughes, 1996; Hsieh, 1996). The use of membranes for combining reaction and separation into a single operation has proven an interesting concept, attracting significant research efforts during the last decade. Ongoing industrial interest in the technology stems from the promise of very compact flow schemes, which can translate to substantial processing cost savings. The interest follows from the earliest developments with inorganic membranes in the fledgling nuclear industry some 50 years ago and the later commercialization successes of polymeric membranes for gas separations. The relative ease with which polymeric membranes have been tailored to be more perm-selective to either reactants or products and the ease with which catalysts can be incorporated into the surface of the membrane have led to increased interest in the potential applications of membranes to a variety of industrial processes. Among these, the immobilization of enzymes on membrane surfaces has led to the commercialization of membrane reactors for a number of food and biotechnology applications. The barrier to application of membranes to higher temperature chemical processes imposed by polymer-based construction is gradually being overcome by the development of inorganic membranes and coating or impregnation of traditional heterogeneous catalysts to the membrane surface. In this chapter, the broad spectrum of catalytic reactions and membranes that have been utilized thus far in reactive separation studies for nonbiological applications is reviewed. Important aspects of process modeling, design, and optimization are discussed and critically evaluated in the context of a concise technical and economic evaluation.

Chapter 7, "Reactive Crystallization," by Vaibhav V. Kelkar, Ketan D. Samant, and Ka M. Ng, reviews reactive separation technologies where one or more of the reaction products appear in the solid phase. Reactive crystallization is a rather common unit operation (Mersmann, 1995; Tavare, 1995). Although poorly documented in the literature, simultaneous chemical reaction and crystallization (or precipitation) are commonly conducted in the batch production of pharmaceuticals and fine chemicals. Reactive crystallization has also been put to work in the large-scale, continuous production of inorganic chemicals such as ammonium sulfate, calcium sulfate, ammonium phosphate, and potassium chloride as well as for waste neutralization with lime (Tavare, 1995; Myerson, 1993). Reactive crystallizations have also been reported for the large-scale production of organic chemicals such as terephthalic acid via liquid-phase air oxidation (Stark, 1979). In this chapter, the combined effects of reaction kinetics on crystal nucleation and growth mechanisms are discussed, modeled, and translated into implications for the design and operation of continuous crystallizers for the production of solid products. The chapter begins with a quick review of the solid–liquid-phase phenomena involved in defining the separation. The interplay among the kinetics of reaction, crystallization nucleation, and crystal growth and their effects on product crystal attributes are then explored. Additionally, the considerations due to heat and mass transfer are also described. The chapter concludes with a discussion of utilizing the current basis of understanding to advance equipment and process design.

1.4 Conclusions

Reactive separations are the subject of increasingly active investigation and development in academia and industry. The coupling of reaction and separations process operations into a single process unit promises improved process design and better process economics in many instances. The prospects for such applications are explored in Chapters 2–7 of this book for the six most common hybridized unit operations: (1) reactive distillation, (2) extraction with reaction, (3) absorption with reaction, (4) adsorption with reaction, (5) reactive membrane separation, and (6) reactive crystallization.

References

Agreda, V.H., L.R. Partin, and W.H. Heise, *Chem. Eng. Progress*, **86**(2), 40, 1990.
Astarita, G., D.W. Savage, and A. Bisio, *Gas Treating with Chemical Solvents*, Wiley & Sons, New York, 1983.
Backhaus, A., U.S. Patent 1,400,849, 1921.
Bland, W.F. and R.L. Davidson, *Petroleum Processing Handbook*, McGraw-Hill, New York, 1967.
Danckwerts, P.V., *Gas–Liquid Reactions*, McGraw-Hill, New York, 1970.
Hilton, C.B., U.S. Patent 4,615,806, 1986.
Hsieh, H.P., *Inorganic Membranes for Separation and Reaction, Membrane Science and Technology, Series 3*, Elsevier, Amsterdam, 1996.
Keyes, D.B., *Ind. Eng. Chem.*, **24**, 1096, 1932.
Kruel, L.U., A. Gorak, and P.I. Barton, *Chem. Eng. Sci.*, **54**, 19, 1999.
Kulprathipanja, S., B.C. Spehlmann, R.R. Willis, J.D. Sherman, and W.A. Leet, U.S. Patent 5,962,735, 1999.

Lander, E.P., J. Nathan Hubbard, and L.A. Smith, *Chem. Eng.*, April 18, 1983, pp. 36–39.

Mersmann, A., Ed., *Crystallization Technology Handbook*, Marcel Dekker, Inc., New York, 1995.

Myerson, A.S., *Handbook of Industrial Crystallization*, Butterworth-Heinemann, Boston, MA, 1993.

Pence, D.T. and W.J. Macek, "Silver Zeolite: Iodide Adsorption Studies," The U.S. Atomic Energy Commission, Idaho Operations Office, under Contract No. AT(10-1)-1230, November, 1970.

Rydberg, J., C. Musikas, and G.R. Choppin, *Principles and Practices of Solvent Extraction*, Marcel Dekker, Inc., New York, 1992.

Scott, K. and R. Hughes, Eds., *Industrial Membrane Separation Technology*, Blackie Academic & Professional, London, 1996.

Stark, L.E., UK Patent 1,555,246, 1979.

Tavare, N.S., *Industrial Crystallization: Process Simulation Analysis and Design*, Plenum Press, New York, 1994.

Terril, D.L., L.F. Silvestre, and M.F. Doherty, *Ind. Eng. Chem. Process Des. Dev.*, **24**, 1062, 1985.

Tien, C., *Adsorption Calculations and Modeling*, Butterworth-Heinemann, Boston, MA, 1994.

Zarzycki, R. and A. Chacuk, *Absorption*, Pergamon Press, Oxford, 1993.

Chapter 2

REACTIVE DISTILLATION

Gavin P. Towler and Stanley J. Frey

2.1 Introduction

The principle of reactive distillation is simple. A distillation column is constructed, a portion of which contains a catalyst for carrying out a desired reaction. The catalyst may be present as a solid-phase packing, or it may be in the same phase as the reacting species. The feeds for the process reaction are fed near to the catalyst section (either above or below, depending on volatility). The reaction occurs, usually mainly in the liquid phase, on the catalyst. The products of the reaction are removed continuously, either in the distillate or the bottoms, while the reagents are kept in the column.

Carrying out reaction and distillation in the same piece of equipment offers several advantages compared to conventional reactor-separator sequences:

- *Shifting of equilibrium.* Removing one or more products from the reaction phase causes the equilibrium to be reestablished at a higher conversion. If the relative volatilities are favorable it may be possible to maintain the reagents in the column and draw off only the products. Even if only one product can be separated from the reaction phase, the increase in conversion still gives a benefit in reduced recycle costs.
- *Reduction in plant cost.* Simplification or elimination of the separation system can lead to significant capital savings.
- *Heat integration benefits.* If the reaction is exothermic, the heat of reaction can be used to provide heat of vaporization and reduce the

reboiler duty. Other heat integration benefits can be obtained through use of intermediate condensers, reboilers and pump-arounds.

- *Avoidance of azeotropes.* Reactive distillation is particularly advantageous when the reactor product is a mixture of species that can form several azeotropes with each other. In such cases a conventional separation scheme would require many distillation columns and use of entrainers to break the azeotropes. Instead, careful choice of reactive distillation conditions can allow the azeotropes to be "reacted away" in a single vessel. Azeotropic behavior can also be exploited to maintain high concentrations of the reagents in the reaction zone of the column, or to separate close-boiling mixtures of a reagent and product.
- *Improved selectivity.* Removing one of the products from the reaction mixture or maintaining a low concentration of one of the reagents can lead to reduction of the rates of side reactions and hence improved selectivity for the desired products.
- *Multifunctional reactor designs.* In some cases several catalyst zones can be included in the same reactive distillation column, allowing more than one reaction function. This is particularly useful when processing an impure feed.

Against these advantages, the disadvantages are:

- *Volatility constraints.* The reagents and products must have suitable volatility to maintain high concentrations of the reagents and low concentrations of products in the reaction zone.
- *Residence time requirement.* If the residence time for the reaction is long, a large column size and large tray hold-ups will be needed and it may be more economic to use a reactor–separator arrangement.
- *Plant scale.* It is difficult to design reactive distillation processes for very large flow rates because of liquid distribution problems, etc.
- *Process conditions.* In some processes the optimum conditions of temperature and pressure for distillation may be far from optimal for reaction and vice versa.

Despite these drawbacks, in some processes the benefits can lead to significant reduction in costs, which has led to considerable interest in this technology since it achieved widespread commercialization in the 1980s. Reactive distillation technology has benefited greatly from the experience gained in its first industrial applications, from improved theoretical understanding and modeling capability, and from advances in catalyst and equipment technology. These subjects will be addressed in the sections that follow, but first the major commercial applications of the technology will be reviewed.

2.2 Industrial Applications

2.2.1 Esterification

Reactive distillation was first considered for esterification processes in a series of patents by Backhaus in the 1920s (Backhaus, 1921a–d, 1922a–e, 1923a,b). In esterification processes the generic reaction is

$$R_1OH + R_2COOH \leftrightarrow R_2COOR_1 + H_2O$$
$$\text{alcohol} + \text{acid} \leftrightarrow \text{ester} + \text{water}$$

(2.1)

Several factors make these systems suitable for reactive distillation. First, mixtures of organic acids, esters, alcohols, and water have a strong tendency to form azeotropes. In any esterification process it is likely that several binary azeotropes will be formed between the reaction feeds and products, and ternary and higher azeotropes may also be possible. Because of this, the separation of the desired product from the reaction mixture is usually an expensive process involving many distillation columns. Second, esterification reactions occur at moderate temperatures in the liquid phase, under conditions that are convenient for distillation. Third, the reactions are equilibrium limited, providing an incentive for removing the products to improve the reactor performance. Finally, because of the presence of azeotropes, it is usually possible to exploit changes in the volatility order in the distillation column and maintain high concentrations of the reagents in the reactive section.

All of these features are illustrated by the methyl acetate process developed by Eastman Chemical in the 1980s (Agreda and Partin, 1984; Agreda et al., 1990). Methyl acetate (MeOAc) is formed by the reaction of methanol (MeOH) with acetic acid (HOAc). Azeotropes form between methyl acetate and methanol, and between methyl acetate and water, and a near azeotrope is formed between acetic acid and water, making separation of the product difficult in conventional processes. Failure to recognize the formation of azeotropes hindered early attempts to use reactive distillation for methyl acetate manufacture (Backhaus, 1921a).

The Eastman Chemical reactive distillation column is illustrated in Figure 2.1. The operation of the process is best understood by considering the function carried out by each section of the column. The top section of the column serves to enrich the product methyl acetate. In this section the only species present at high concentrations are methyl acetate and acetic

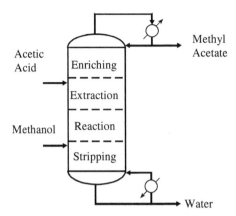

Figure 2.1: The Eastman Chemical methyl acetate process.

acid. No azeotropes form between these species, and this section therefore acts as a conventional rectifier. The acetic acid feed enters the column below the enriching section and flows down through the extraction section. In the extraction section of the column, acetic acid acts as an entrainer for any water that has been carried up in the vapor and breaks the methyl acetate/water azeotrope by extractive distillation.

The acetic acid then enters the reactive stages of the column, where it meets methanol and undergoes esterification. Sulfuric acid is used as a catalyst for the reaction, and the reactive stages are designed so that conversion of acetic acid is complete. The methyl acetate that is produced in the reactive stages vaporizes, removing the heat of reaction and shifting the reaction equilibrium to higher conversion. Finally, in the bottom section of the column any unreacted methanol and acetic acid are stripped from the water byproduct.

Several practical problems had to be overcome in the commercial development of this process, and these are described by Agreda et al. (1990). The identification of a liquid-phase catalyst simplified the process equipment, and it was found that a side draw and stripping column was needed to remove small amounts of intermediate boiling impurities. After testing at the bench and pilot-plant scales in an accelerated program, Eastman Chemical commissioned a 23 metric tons per hour commercial unit in May, 1983. This plant underwent a series of minor modifications to remove bottlenecks and optimize its operation, and by 1990 Eastman Chemical was able to report that it was operating at 125% of design capacity. The Eastman methyl acetate process has since served as a case example in many academic studies of reactive distillation, because the absence of solid-phase catalyst makes it a relatively simple system to study.

2.2.2 Etherification

The biggest commercial application of reactive distillation has been in the production of ethers for blending into gasoline. The 1990 amendments to the U.S. Clean Air Act mandated reformulated gasoline for regions of the United States that were failing to meet air quality standards. The requirements for reformulated gasoline included restrictions on the maximum content of aromatic hydrocarbons and the minimum concentration of oxygenated species in the gasoline. Ethers are a very attractive blending component for reformulated gasoline. As they have high octane numbers that compensate for the reduction in aromatic compounds, they can satisfy the oxygen requirement, and they do not significantly increase volatile organic compound (VOC) emissions. Structures and blending properties of some of the ethers that are used in gasoline are shown in Figure 2.2.

The ethers used in gasoline production are made by reaction of an alcohol such as methanol or ethanol with a tertiary olefin such as isobutene or isopentene, according to the generic equation

$$RC(CH_3){=}CH_2 + R'OH \leftrightarrow RC(CH_3)_2OR'$$
$$\text{tertiary olefin} + \text{alcohol} \leftrightarrow \text{ether} \tag{2.2}$$

This reaction is carried out at moderate temperatures in the range 60 to 100°C and at pressures below 1.5 MPa. The process is limited by the equilibrium of reaction 2.2. Conventional processes require a complex system of distillation columns to recover the unconverted olefin and alcohol for recycling. This separation is usually complicated by the presence of azeotropes between the alcohol/ether and alcohol/olefin.

The use of reactive distillation allows a much simpler process, in which the reagents (or azeotropic mixtures containing the reagents) are stripped from the product in the lower section of a reactive column and the ether is withdrawn as bottom product. This process was initially developed by Chemical Research and Licensing Ltd. for production of methyl tertiary butyl ether (MTBE). Following successful demonstration of the reactive distillation route to MTBE (Smith, 1981b; Lander et al., 1983), similar processes were developed for the other ethers used in blending. Several companies now license reactive distillation processes for manufacture of ethers, with the main difference between these processes being in the type of column internals.

One of the most important advantages of reactive distillation for ethers production is that the process does not require a high-purity feed of iso-

Methyl Tertiary Butyl Ether (MTBE)

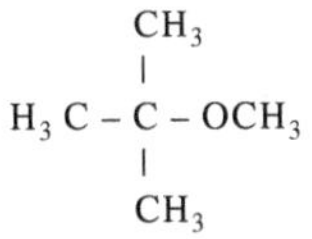

$$CH_3$$
$$|$$
$$H_3 C - C - OCH_3$$
$$|$$
$$CH_3$$

Blending Octane (R + M)/2 = 109
Blending RVP* = 8 psig

Ethyl Tertiary Butyl Ether (ETBE)

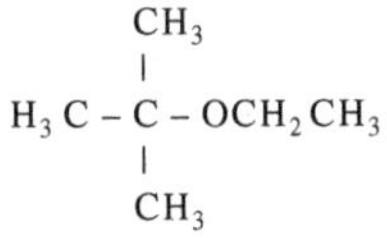

$$CH_3$$
$$|$$
$$H_3 C - C - OCH_2 CH_3$$
$$|$$
$$CH_3$$

Blending Octane (R + M)/2 = 110
Blending RVP* = 4 psig

Tertiary Amyl Methyl Ether (TAME)

$$CH_3$$
$$|$$
$$H_3 C - C - OCH_3$$
$$|$$
$$CH_2 H_5$$

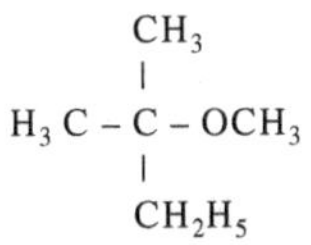

Blending Octane (R + M)/2 = 105
Blending RVP* = 3 psig

*Reid Vapor Pressure

Figure 2.2: Structures and properties of ethers used in gasoline blending.

olefin. This is of great benefit in oil refineries, where the iso-olefin is typically available in a mixture with other olefins and paraffins of the same boiling range. The reactive distillation column can be designed so that full conversion of the iso-olefin is achieved while all the other components of the olefinic feed pass through the process and emerge in the distillate or bottoms for further processing or product blending. The reactive distillation can also be designed to accomplish other reactions that are necessary to treat a mixed feedstock. The CDTech tertiary amyl methyl ether (TAME) process described by Hickey and Adams (1994) illustrates this.

In the Hickey and Adams process, methanol is reacted with isoamylenes to form TAME. The isoamylenes are fed as a light naphtha stream, which can also contain diolefins, acetylenes, other C_5 compounds and some C_6 compounds. The concentration of isoamylenes is usually quite low, typically below 15 mol%. Depending on the processes that have been used to produce this stream there may be significant concentrations of mercaptan compounds present. Diolefins and acetylenes inhibit the etherification reaction, while mercaptans must be removed to meet fuel specifications, so it is advantageous if these compounds can be removed during the reactive distillation.

Figure 2.3 is a schematic of the process. Once again, the process is best understood by considering the function of each section of the column. The top section is a rectifier that separates hydrogen, light ends, and unreacted C_5 compounds, and provides a reflux rich in C_5 olefins to the stages below. The etherification reaction takes place in the upper reaction section of the column. This section contains a catalyst for etherification, such as an acid cation exchange resin. The catalyst can be loaded into the column in a number of different ways, as discussed in Section 2.6.

The TAME that is formed is less volatile than the reagents and so flows back down the column. The lower reaction section is used to accomplish several reactions that improve the quality of the feed. The catalyst in this section is a hydrogenation catalyst such as palladium on alumina. Hydrogen is fed with the mixed C_5/C_6 stream, and in the presence of the hydrogenation catalyst diolefins and acetylenes are saturated to mono-olefins. Isomerization of olefins also occurs, increasing the concentration of isopentene that can be converted to TAME. The hydrogenation section also removes mercaptans from the feed by conversion to H_2S. A low-boiling azeotrope between methanol and isopentene is formed in this section, which improves separation of isopentene and methanol from the heavier species and sends the reagents up into the upper reaction zone. Finally,

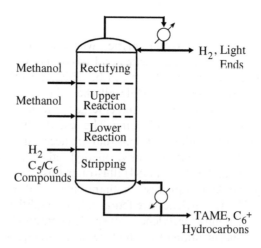

Figure 2.3: Hickey and Adams TAME process.

the stripping section at the base of the column removes methanol and C_5 hydrocarbons from the C_6 hydrocarbons and TAME product.

This process shows that reactive distillation is not restricted to treating pure feeds or to processing the relatively small flow rates usually encountered in chemical plants. In all, over 140 reactive distillation ether plants have been commissioned since 1985, with a typical output of 1500 barrels per day (roughly 200 metric tons per day). Reactive distillation is now the most common process used for manufacture of ethers.

2.3 Theory

The strong industrial interest in reactive distillation in the past 10 years has led to many attempts to deepen theoretical understanding of reactive distillation columns and to develop methods for their design. This section provides an introduction to some important equilibrium properties of reactive distillation and introduces the fundamentals of rate-based modeling of reactive distillation processes.

2.3.1 Equilibrium Behavior

Although the modeling of reactive distillation necessarily must consider rates of reaction, useful insights into the limiting behavior of the process can be gained by considering the behavior of the system at equilibrium.

Venimadhavan et al. (1994) showed that the equilibrium behavior of reactive distillation processes varies between two extremes, corresponding to chemical equilibrium control and phase equilibrium control. This can be illustrated by a simple example. Consider the reaction

$$A + B \leftrightarrow 2C \qquad (2.3)$$

This reaction has only three components, and so is not truly representative of reactive distillation processes, but it allows us to examine a two-dimensional plot of the composition space and to visualize the effect of phase or reaction equilibrium.

The equilibrium constant of the reaction is

$$K = \frac{x_C^2}{x_A x_B} \qquad (2.4)$$

where x_i is the mole fraction of component i, and K can be found from the change in Gibbs free energy of reaction.

If the value of the equilibrium constant is known, then a locus can be plotted through all the possible mixtures in the ABC composition space that can satisfy equation 2.4. Figure 2.4(a) shows such a plot. Any starting composition will react according to equation 2.3 until it reaches a condition at which its composition satisfies equation 2.4. At this point chemical equilibrium is satisfied and there will be no further reaction. Hence, any reaction process can be plotted in the triangular diagram as a vector that follows equation 2.3 from a given initial composition and terminates at the line that satisfies equation 2.4, as shown in Figure 2.4(a) for three initial compositions.

Figure 2.4(b) shows the same diagram with chemical equilibrium lines plotted for several values of K. At low values of K the chemical equilibrium curve lies close to the $A–B$ edge of the space. This is to be expected, as at low values of K little of the feed mixture is converted. At the opposite limit of high values of K, the chemical equilibrium curve lies close to the $A–C$ and $B–C$ edges. This behavior is again expected, as the chemical equilibrium strongly favors formation of C. In this case, the feed component that is initially least abundant is almost completely consumed at chemical equilibrium. The equilibrium product therefore essentially contains only product C and the balance of the other feed component.

The composition diagram showing the chemical equilibrium behavior of the mixture can be contrasted with the diagram that shows the phase equilibrium behavior. For distillation processes the relevant diagram to display phase equilibrium behavior is the residue curve map or distillation line map. These diagrams show the composition profiles at total reflux for packed columns and staged columns respectively. Each residue curve (or distillation line) is marked with an arrow that indicates the direction of increasing temperature. Widagdo and Seider (1996) give an excellent review of the use of these plots. Figure 2.5 shows a residue curve map for a system that exhibits low-boiling binary azeotropes between components $A–B$ and $A–C$.

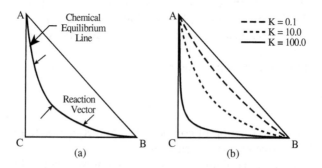

Figure 2.4:   Chemical equilibrium ternary diagrams.

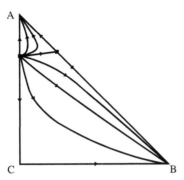

Figure 2.5: Distillation residue curve map for a system with two binary azeotropes.

A feature of residue curve maps is that the residue curves begin and end at stationary points of the differential equation from which the residue curves are calculated. These stationary points correspond to the pure components and the azeotropes. Three kinds of stationary point exist:

Unstable node—all residue curves point outwards.
Stable node—all residue curves point inwards.
Saddle point—some residue curves point out and some point in.

For example, in Figure 2.5, pure components *A* and *B* are stable nodes, the *A–C* azeotrope is an unstable node, and pure component *C* and the *A–B* azeotrope are saddle points. Doherty and Perkins (1979) developed a mathematical relation between the number of stationary points of each type. The stationary points are important in distillation design, as the nature of the stationary point can tell us whether a certain azeotrope or pure component will be easy or difficult to achieve by distillation.

Reactive distillation systems can also contain azeotropes that act as stationary points. The existence of such reactive azeotropes was suggested by Barbosa and Doherty (1987) and proved in an elegant set of experiments reported by Song et al. (1997). Ung and Doherty (1995) describe the necessary and sufficient conditions for a reactive azeotrope. If a reactive azeotrope occurs, then a constant boiling mixture is formed with constant, but not equal, vapor and liquid compositions. This can happen if the rate of reaction and the rate of vaporization balance such that vaporization occurs at constant composition.

The equilibrium behavior of most reactive distillation systems will be intermediate between the phase equilibrium and reaction equilibrium cases. Venimadhavan et al. (1994) suggested using the Damköhler number to model this variation. The Damköhler number, Da, is defined as

$$Da = \frac{H_0 k_1}{V} \qquad (2.5)$$

where H_0 is the liquid hold-up (mol); k_1 is a pseudo-first-order rate constant (s^{-1}), and V is the vapor rate (mol s^{-1}).

The Damköhler number is the ratio of a characteristic liquid residence time (H_0/V) to the characteristic reaction time $(1/k_1)$. For low values of Da, Da < 0.5, the reaction rate is slow relative to the residence time available, and the system is dominated by phase equilibrium. For large values of Da, Da > 10, the reaction rate is fast and chemical equilibrium is approached on the reactive stages. If the Damköhler number lies between these values then neither phase equilibrium nor chemical equilibrium is controlling and a rate-based model must be used, as described in the following section.

It is important to understand the existence of reactive azeotropes, as the formation of a reactive azeotrope could in principle limit the performance of a reactive distillation process. In practice, this requires exactly the right balance between the rate of reaction and the rate of vaporization, and so it is likely to be a transient phenomenon in an industrial process. Nonetheless, reactive distillation columns commonly exhibit sections in which the composition changes only slightly from stage to stage, due to proximity of the composition to a stationary point. Venimadhavan et al. (1994) present a detailed analysis of this situation and show how the location of the stationary points varies between the phase equilibrium and chemical equilibrium limits as the Damköhler number is increased.

2.3.2 Kinetically Limited Behavior

Given the difficulties in modeling reactive distillation as an equilibrium-stage process, most distillation experts might be excused for thinking that nonequilibrium modeling would be an arduous task. A nonequilibrium model of reactive distillation must consider:

- Phase equilibrium
- Rate of mass transfer between the vapor and liquid phases
- Mass transfer within the reacting (usually liquid) phase
- Rate of reaction
- Catalyst activity and possibly also mass transfer within the catalyst
- Mass transfer of products within the liquid phase

This seems to add considerable complexity to the problem, but in fact well-tested modeling methods with decades of industrial acceptance are available from the field of reactive absorption.

These models start from the two-film theory of absorption. Figure 2.6 shows a diagram of the vapor–liquid interface, where z is the distance measured from the vapor into the liquid. It is assumed that the vapor and liquid are in equilibrium at the interface and that all the resistance to mass transfer occurs across a boundary layer of thickness δ in each phase.

To simplify the analysis, consider a case in which component A diffuses from the vapor phase and reacts with nonvolatile component B in the liquid phase according to

$$A + B \rightarrow \text{products} \tag{2.6}$$

Considering the vapor-phase mass transfer, the flux of A through the vapor-phase boundary layer, N_{AG}, can be written as

$$N_{AG} = \frac{D_{AG}}{\delta_G}(p_A - p_{Ai}) = k_g(p_A - p_{Ai}) \tag{2.7}$$

where D_{AG} is the vapor-phase diffusivity of A; δ_G is the vapor-phase boundary layer thickness; p_A is the partial pressure of A in the bulk vapor; p_{Ai} is the partial pressure of A at the interface; and k_g is the vapor-phase mass transfer coefficient.

Similarly, the flux of A through the liquid-phase boundary layer, N_{AL}, is

$$N_{AL} = \frac{D_{AL}}{\delta_L}(C_{Ai} - C_{A,bulk}) = k_L(C_{Ai} - C_{A,bulk}) \tag{2.8}$$

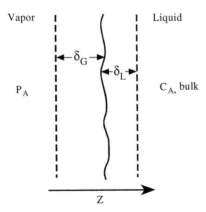

Figure 2.6:   Two-film model of the vapor–liquid interface.

where D_{AL} is the liquid-phase diffusivity of A; δ_L is the liquid-phase boundary layer thickness; $C_{A,bulk}$ is the concentration of A in the bulk liquid; C_{Ai} is the concentration of A at the interface; and k_L is the liquid-phase mass transfer coefficient.

At the interface, vapor–liquid equilibrium is assumed, which can be expressed as

$$p_{Ai} = HC_{Ai} \tag{2.9}$$

where H is the Henry's law constant for component A.

The rate of reaction, r_A, will generally be a function of the concentrations of both reacting species

$$r_A = r_A(C_A, C_B) \tag{2.10}$$

where C_B is the concentration of component B in the liquid phase.

If the concentration of B is large, or if the reaction per column stage is small, then we can make the approximation that C_B is constant, in which case a pseudo-first-order equation can be written

$$r_A = k_1 C_A \tag{2.11}$$

where k_1 is the pseudo-first-order rate constant.

If the reaction is slow, then any reaction that occurs in the liquid film can be neglected, in which case the rate of mass transport through the liquid film and the rate of reaction in the bulk liquid can be equated:

$$N_A a = k_L a(C_{Ai} - C_{A,bulk}) = k_1 C_{A,bulk} \tag{2.12}$$

where a is the surface area per unit volume of the liquid phase (typically this is of the order $1000 \text{ m}^2 \text{ m}^{-3}$ for most vapor–liquid contacting devices).

This equation can be rearranged to give

$$C_{A,bulk} = C_{Ai} \frac{ak_L}{k_1 + ak_L} \tag{2.13}$$

hence

$$N_A a = k_L a C_{Ai} \left(\frac{k_1}{k_1 + ak_L} \right) \tag{2.14}$$

We can distinguish two cases that result from this equation. If the reaction rate constant is much less than the product ak_L, then the overall rate of transport is given by

$$N_A a \approx k_1 C_{Ai} \tag{2.15}$$

Under these conditions the bulk concentration is approximately the same as the interfacial concentration, as shown in Figure 2.7(a). This is known as the

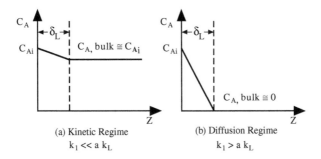

Figure 2.7:   Concentration profiles in the slow regimes.

kinetic regime, as the overall rate is dominated by the kinetic rate of reaction.

Conversely, if the reaction rate constant is much greater than the product ak_L, then the overall rate of transport is given by

$$N_A a \approx k_L a C_{Ai} \tag{2.16}$$

This equation is the same as that which would be obtained if only physical absorption were occurring, with a bulk concentration of A approaching zero. This condition is shown in Figure 2.7(b). In this case, it can be seen that the controlling process on the liquid side is mass transfer across the film, hence, this is termed the diffusion regime. The slow kinetic and slow diffusion regimes are very important in reactive distillation, as most reactive distillation processes maintain significant concentrations of all reacting species in the liquid phase, and operate in the slow kinetic regime or between the slow kinetic and slow diffusion regimes.

The condition that determines whether the process is in the slow regimes was stated above, and it is that the rate of reaction in the liquid film is negligible. This condition can be quantified by finding the conditions for which the rate of transport through the film is much greater than the rate of reaction within the film.

Rate of transport through film (mol s^{-1}) $\gg$

rate of reaction in film (mol s^{-1})

$$k_L a(C_{Ai} - C_{A,bulk}) \gg (a\delta)k_1 C_{A,film} \tag{2.17}$$

where $C_{A,film}$ is the average concentration of A in the film.

Since the average concentration of A in the film is less than the concentration at the interface and the concentration of A in the bulk liquid approaches zero as the reaction rate increases, equation 2.17 can be approximated as

$$k_L a C_{Ai} \gg a \delta k_1 C_{Ai} \tag{2.18}$$

hence

$$\frac{D_A k_1}{k_L^2} \ll 1 \quad \text{since} \quad k_L = \frac{D_A}{\delta_L} \tag{2.19}$$

This is used to define the Hatta number,

$$\text{Ha} = \frac{\sqrt{D_A k_1}}{k_L} \tag{2.20}$$

So the condition for either of the slow regimes is

$$\text{Ha}^2 \ll 1 \tag{2.21}$$

If reaction in the liquid-side film is significant, then the concentration profile in the film becomes steeper with increasing rate of reaction, as shown in Figure 2.8. This is known as the fast regime, and in this case, the rate is written as

$$N_A a = k_L a C_{Ai} E \tag{2.22}$$

where E is the enhancement factor, defined as the rate of absorption with reaction divided by the rate of physical absorption.

It can be shown that

$$E = \text{Ha coth (Ha)} \tag{2.23}$$

and for Ha > 3.0,

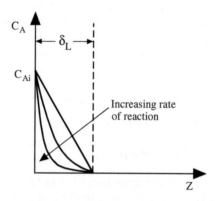

Figure 2.8: Concentration profiles in the fast regime.

$$E \approx \text{Ha} \qquad (2.24)$$

This is valid until diffusion of component B begins to limit the rate, which is known as the instantaneous reaction regime. The instantaneous regime is important in absorption of acid gases but is unlikely to occur in reactive distillation processes.

Calculation of the Hatta number therefore allows us to determine which rate processes govern the overall reaction rate in the reactive distillation process. The practical application of this analysis to the modeling and design of reactive distillation processes will be described in the section below.

2.4 Modeling and Design

Development of realistic process models is a key step in evaluating the feasibility of a proposed reactive distillation process. Much of this modeling can now be carried out using commercial simulation software provided by vendors such as Aspentech, Hyprotech, and Simulation Sciences. A successful modeling effort requires a full understanding of both the equilibrium and the kinetic limits on the process.

2.4.1 Equilibrium Modeling

Modeling the equilibrium nature of the process must begin with determination of the chemical equilibrium constant and the phase equilibrium behavior of the nonreacting system. The presence of azeotropes can usually be determined by looking at binary, ternary or quaternary data for mixtures of the feeds and products in the absence of catalyst. The phase equilibrium behavior is also important for any nonreacting stages of the reactive distillation column, for example rectifying or stripping stages above or below the reactive section of the column.

The Damköhler number and chemical equilibrium constant can be used to make a preliminary screening of the suitability of the process for reactive distillation, as proposed by Venimadhavan et al. (1994). Table 2.1, adapted from their paper, shows the processing impact of different combinations of Da and K.

If reactive distillation is a viable process option, then the Damköhler number can be used to determine whether the process will be controlled by phase equilibrium or by chemical equilibrium, as described in Section 2.3.1.

Table 2.1: Processing Implications of Da and *K*.

Da	*K*	Processing Impact
Low	Low	Slow forward reaction and fast reverse reaction. Essentially, no product formation occurs and the process is very difficult to carry out in any vessel. Reactive distillation will usually not offer any advantage over an optimally designed reactor because of the large hold-up required.
Low	High	Slow forward reaction, but slower reverse reaction. In this case, the product formation is high. Reactive distillation can give benefits as long as required tray hold-up is not too large.
High	Low	Fast product formation but faster reverse reaction, so little product formation. Reactive distillation can be advantageous if products can be removed from reactive zone quickly enough to prevent reverse reaction.
High	High	Fast forward reaction and slower reverse reaction. Essentially, this is an irreversible reaction, which could be carried out in a simple reactor. Reactive distillation is usually not justified unless it is used to obtain conditions that minimize competing reactions.

- Da < 0.5: *phase equilibrium control.* If phase equilibrium is controlling, then the reactive stages will be strongly influenced by azeotropes that are present in the nonreacting mixture. For low values of Da a large residence time is necessary. Rate-based models must be used for column design in this regime, as phase equilibrium models are unable to account for the appearance of the products.
- Da > 10.0: *chemical equilibrium control.* In this case, the reactive stages can be assumed to be at chemical equilibrium. Each reactive stage can be modeled as an equilibrium chemical reactor with vapor and liquid products. Several commercial simulators now have the ability to model such stages.
- $0.5 <$ Da < 10.0: *rate limited.* In the intermediate range of Da, rate-based models are needed as both kinetic and phase equilibrium effects are important.

Note that since the Damköhler number depends on the tray hold-up (or stage hold-up for a packed column) and the vapor rate, the design engineer has some scope for varying this parameter by changing the column design. There is also considerable scope for varying the Damköhler number by changing the rate constant. The rate constant is easily changed by changing the operating temperature of the process through small changes in column pressure. The range of values of Da that can be achieved will be limited by practical considerations such as choice of column internals, flooding, entrainment, etc.

2.4.2 Rate-Limited Modeling

If the Damköhler number is less than 10.0, then the process will not be chemical equilibrium controlled and rates of reaction and mass transfer must be taken into account. This need not require a complex rate-based modeling effort if the results presented in Section 2.3.2 are used properly. The first step is to calculate a pseudo-first-order rate constant for the reaction and to estimate the diffusivity and mass transfer coefficient in the liquid phase for the reagent species that is expected to have the lowest liquid phase concentration on the reacting stages. (If both reacting species are expected to have high concentrations then the process is already known to operate in the slow kinetic regime.)

With this information the Hatta number can be calculated and the value of ak_L can be estimated. Using these values, we can determine which processes are rate controlling and hence, which design method and modeling approach to use.

- Ha $\ll$ 1: *slow regimes.*
- $k_1 \ll ak_L$: *slow kinetic regime.* This will often be the case for reactive distillation processes, as the concentration of all reacting species in the liquid phase is often significant. In this regime the rate is most sensitive to the intrinsic rate constant and is insensitive to the surface area per unit volume of column. Hence better catalyst improves the design, but improved column packings do not. The design of the reactive stages will be dominated by providing sufficient liquid hold-up to achieve the required residence time. A rate-based model can be developed that assumes phase equilibrium and reaction kinetics only, as the rate of mass transfer does not influence the overall rate and can be neglected.
- $k_1 \gg ak_L$: *slow diffusion regime.* In this case, mass transfer controls the overall rate. The process will benefit from improvements in the vapor–liquid contacting such as use of advanced column packings, and there is no benefit from improvements in the catalyst. A rate-based model can assume that the reaction is in equilibrium in the liquid phase and need only consider the rate of mass transfer between phases.
- $k_1 \approx ak_L$: *slow mixed regime.* If the kinetic and diffusion terms are approximately equal, then neither dominates the overall rate, and the rate can be found from equation 2.14. In this case, the rate-

based model must consider both reaction and mass transfer. The process will be enhanced by improvements in either vapor–liquid contacting or catalyst.

- Ha $\gtrsim 1$: *fast regime*. In the fast regime, the overall rate depends mainly on the enhancement factor. For Ha > 3.0, the overall rate is independent of the mass transfer coefficient, but proportional to the interfacial area, a, and the square root of the reaction rate constant. Improvements in vapor–liquid contacting are therefore effective if they increase the interfacial area. Improvements in catalyst performance give improved performance, but are less effective than might be expected because of the half-order proportionality. Rate-based models must be used in this regime, and care must be taken to ensure that the model considers reaction in the film as well as in the bulk liquid, otherwise the results will erroneously default to the slow diffusion model.

Once again, the designer can influence the behavior of the process by varying the temperature of the column. For rate-limited designs, however, the scope for variation is limited, as D_A and k_L are also temperature dependent, though not usually as strongly as k_1.

Once the conceptual design of the process is complete, and has been confirmed by modeling and simulation, the next step is to consider the practical aspects of designing the reactive distillation column.

2.5 Practical Design Considerations

Unlike most of the other reactive separation processes presented in this book, reactive distillation has been used widely on a commercial scale. Because of its relatively mature state of development, many solutions to practical problems in scaling up reactive distillation processes have already been implemented. These practical problems for the most part arise from scaling up to commercial production levels and have been addressed by development of novel mechanical devices.

The primary practical issues in implementing a large-scale reactive distillation application are to determine how to:

- Install, hold, and remove the catalyst
- Ensure good contacting of the reactive phase with the catalyst
- Ensure good vapor–liquid contacting throughout the reactive zone
- Avoid excessive pressure drop through the reactive zone
- Allow sufficient hold-up in the reactive section, *and*
- Allow for catalyst deactivation

The main focus of this section will be the case where solid catalysts are required to produce the desired reaction, which has been necessary for all of the industrial applications of reactive distillation, with the exception of the Eastman methyl acetate process.

Each of these issues will be discussed in more detail and then industrially proven designs and other design concepts that have been forwarded to address these issues will be reviewed.

2.5.1 Installation, Containment, and Removal of the Catalyst

Distillation columns in bulk production of chemicals and fuels are usually quite large with internal access restricted by the size of small manways, 45–60 cm in diameter. The large amount of catalyst and mechanical equipment that must be installed in the vessel through these small manways must be considered in the mechanical design. It is highly desirable to allow easy installation and removal of the reactive distillation equipment and catalyst. The constraints imposed by the design of existing columns are paramount, as many of the reactive distillation applications in industry are installed as revamps to existing units. Even if new distillation column construction is feasible, permanently installed reactive distillation equipment reduces the scope of application of that particular column. Thus, the use of devices that can be installed and removed through manways has been the common practice.

After the catalyst has deactivated it must be removed from the column or regenerated in place. In conventional reactor-only technology, solid catalysts are commonly regenerated by combustion of the catalyst surface foulants. In reactive distillation, the metallurgy required to perform catalyst regeneration by combustion may be too costly because of the large size of the distillation column. As an alternative, it may be possible to regenerate the catalyst by passing some reactivating material over it. However, this is not currently practiced because of the large volume of hazardous waste usually produced by such regeneration methods. Given these in-situ regeneration difficulties, removal of the catalyst following deactivation has been the common practice in industrial applications of reactive distillation. To allow catalyst removal, the reactive distillation device must allow for the reaction section to be well purged of toxic compounds that would make entry to the column hazardous.

Adding a function to remove catalyst without personnel entry or even move the catalyst to a regeneration section external to the distillation column while production continues is of great interest. Development of a device that allows removal of solid catalyst is a subject of great interest. Reactive distillation is often passed over as a processing option because the catalyst life would require frequent distillation column shutdowns or the risk of a shutdown forced by a catastrophic catalyst poisoning is too great. A reactive distillation device that allowed onstream removal of catalyst would answer this concern and open the application of reactive distillation to other chemicals.

2.5.2 Design for Good Reactive Phase Contact with the Catalyst

Efficient contacting of the reactive phase with the catalyst is necessary to minimize the required catalyst volume, avoid radial concentration gradients, prevent reactor hot spots, allow even catalyst aging, and maximize conversion. For systems with solid catalysts, the reactive phase is predominantly liquid, as almost all the solid surfaces in a distillation column are wetted. Thus the problem of reactive phase contacting with the catalyst is reduced to a question of liquid-phase contacting with solids.

Designing for even distribution of liquid over a solid surface starts with ensuring even liquid distribution over the cross-section of the column. Many distributors are used commercially for this purpose. Some have been specifically designed for catalytic distillation, for example, that proposed by Hearn et al. (1996).

Given that the liquid begins its descent through the catalyst section evenly, the catalyst-containing equipment must be designed so that the liquid remains evenly distributed, is radially well mixed, and is not allowed to bypass around the catalyst but instead is forced into intimate contact with it. Maintenance of an even flow distribution and good radial mixing are typically ensured by a combination of means. These include using frequent redistributors, running partially flooded beds, producing criss-crossing mixing patterns by alternating the arrangement of the catalyst-containing device, and using trays that are designed to impose a horizontal velocity vector on the rising vapor, thereby thoroughly mixing the liquid on the tray. This last option also provides for intimate catalyst–liquid contacting. In designs where the catalyst is contained from movement and a flooded zone is not desired, the liquid–catalyst contacting is accomplished by forcing the liquid through tortuous paths as it falls down the catalyst section, leaving flow through the catalyst as the direct vertical path. Examples of the application of these concepts will be seen in the review of commercially available equipment that is given in Section 2.6.

2.5.3 Design for Liquid–Vapor Contacting Through the Reactive Zone

Depending upon the need for simultaneous reaction and distillation and the regime in which the reactive stages operate, the contacting efficiency of liquid and vapor in the reactive section may or may not be important. (Simultaneous reaction and distillation is defined as significant amounts of each occurring in the same column height.) If the reaction rate is fast and the reaction is equilibrium limited, then the required size of the reactive zone is strongly influenced by the effectiveness of the vapor–liquid contacting and height equivalent of a theoretical vapor–liquid mass transfer plate (HETP). In this case the reaction rate is primarily a function of the rate of separation of products from reactants; devices that have minimal vertical back-mixing of the liquid and vapor streams but good horizontal mixing of the liquid and vapor streams are preferred.

Often the reaction kinetics or catalyst–pore diffusion limits the reaction rate. In this case, simultaneous reaction and distillation is not required and a flooded back-mixed bed of catalyst in the distillation column may be adequate, perhaps even preferred, if it allows for the addition of more of the reaction-limiting catalyst. The operation is really reaction followed by distillation with little or any separation occurring on the reaction stages. Economic designs can still be realized by stacking several reactors on top of each other in the same column and placing separation stages either above, below, or between them.

Mechanical designs for good vapor–liquid contacting generally resemble devices that are already used for distillation, as vapor–liquid mass transfer is the primary consideration when designing for distillation column internals. These devices include structured packing, random packing, and distillation trays.

2.5.4 Design for Proper Pressure Drop Through the Reactive Zone

Pressure drop through the reactive zone may or may not be important to the overall design, depending upon the static pressure of the reactive distillation operation and the mass fluxes through the reactive zone. In general, the importance of having a device with low pressure drop increases as the static pressure decreases and the mass fluxes increase. The primary considerations in designing for reactive zone pressure drop are providing for enough mass flow capacity and allowing a high enough pressure in the overhead section to economically condense the vapor. Pressure drop has little direct influence on the reaction.

Designs for low pressure drop generally provide allowance for consistent open flow paths for the rising vapor. If the rising vapor is forced through the liquid then frothing is promoted and the resistance to downward liquid flow increases; thus, liquid builds up in the column and "flooding" occurs.

2.5.5   Design for Proper Liquid Hold-up

While flooding is usually not preferred in the reactive section, a significant amount of liquid hold-up must be maintained to give the liquid some residence time with the catalyst. To slow the liquid from "free fall" through the reactive zone, the reactive distillation device may include some feature or added element to make the falling liquid flow in inclined paths rather than directly vertical. This gives the liquid flow stream longer path lengths in the reactive zone without increasing the height of the reactive zone. The flatter the flow path, the longer the residence time. The amount of liquid hold-up must be optimized against flooding considerations, as mentioned in the previous paragraph. The greater the horizontal component of the liquid flow velocity, the earlier the reactive zone will flood.

2.5.6   Design for Catalyst Deactivation

As mentioned in Section 2.5.1, development of a device that allows online removal of the catalyst is very desirable; however, a device to conveniently accomplish this has not yet been commercialized. Instead, catalyst deactivation is typically compensated at the design stage by assuming a deactivation level and adding an additional amount of catalyst to allow for the amount that is deactivated. The drawbacks to this method include the promotion of unwanted consecutive reactions, added catalyst cost, and additional vessel cost if the added catalyst forces an increase in the diameter or height of the distillation column.

Besides adding excess catalyst, some allowance for increasing reflux or column pressure can also be included in the design so that the reaction severity can be increased throughout the life of the catalyst. Increasing pressure effectively speeds the reaction by increasing the temperature and the extent of reaction (so long as the reaction is not exothermic and severely equilibrium limited). Increasing the reflux can enhance conversion in equilibrium-limited reactions by giving steeper concentration profiles in the reactive zone and thereby increasing the concentration driving force for the reaction. Increased reflux can also effectively increase the number of passes through the reactive zone, enhancing conversion for kinetically limited reactions.

There has been much development of new or modified distillation equipment for reactive distillation to produce physical answers to the above questions. Depending upon the emphasis the developer places on the relative importance of each of these issues, the resulting devices can appear quite different from one another. Of course the overriding issue is that of cost. The new equipment cost must not nullify the economic benefits gained by using a reactive distillation system rather than a conventional reaction–distillation–recycle flow scheme. The following section reviews some of the equipment ideas that have been generated recently. In each example it is important to consider how the device would perform in regard to the criteria described above.

2.6 Commercially Proven Equipment Technology

While many concepts for reactive distillation equipment have been patented, only two are in wide commercial use today. These equipment designs were both developed by companies that license process technology to the oil and chemicals industries and do not themselves use the equipment for the manufacture of chemicals or fuels.

2.6.1 Chemical Research & Licensing (CR&L) Catalyst Bales

In 1980, Chemical Research & Licensing (at the time a new research and development company) licensed the first widely publicized commercial application of reactive distillation for the production of methyl *tert*-butyl ether (MTBE), as described in Section 2.2.2. Chemical Research & Licensing developed a method to hold catalyst in the distillation column using fiberglass cloth (Arganbright et al., 1986). Pockets are sewn into a folded cloth and then solid catalyst is loaded into the pockets. The pockets are sewn shut after loading the catalyst and the resulting belt or "catalyst quilt" is rolled with alternating layers of steel mesh to form a cylinder or "catalyst bale" as shown in Figure 2.9. The steel mesh creates void volume to allow for vapor traffic and vapor–liquid contacting.

Scores of these bales are installed in the reactive zone of a typical commercial distillation column. It is important that the bales are properly packed into the column to avoid vapor channeling. Bales are piled on top of each other to give the required height necessary to achieve the desired extent of reaction. When the catalyst is spent, the column is shut down and the bales are manually removed and replaced with bales containing fresh catalyst.

Figure 2.9:   Chemical Research & Licensing catalyst bales.

Chemical Research & Licensing has accumulated much experience with the catalyst bales in the commercial application of reactive distillation to etherification, hydrogenation, and alkylation of aromatic compounds (Barchas et al., 1993; Shoemaker and Jones, 1997). Improvements to the catalyst bale concept have been made over the years (Johnson, 1993; Crossland et al., 1995). In addition, operational experience has given rise to new design ideas that compensate for catalyst deactivation by increasing the liquid hold-up around the catalyst bales (Smith, 1992). The mass transfer characteristics of the bales were recently evaluated by the Separations Research Institute (SRI) (Subawalla et al., 1997). Plots of mass transfer efficiency and capacity are shown in Figure 2.10.

The design of the catalyst bale allows for a wide variety of conventionally formed solid catalysts to be loaded into the same generic structure. This design allows application to a wide variety of catalysts and reactions with little or no change in the catalyst-containing device. Depending upon the amount of catalyst needed, the catalyst can be packed less densely by substituting inert material for a portion of the catalyst.

2.6.2   Koch-Glitsch, Inc. Catalyst-Containing Structured Packing

In the early 1990s, the Koch Engineering Company (now Koch-Glitsch, Inc.) took the basic design of structured distillation packing and modified it so that it could hold catalyst. Typically, structured distillation packing consists of corrugated thin sheet metal or wire gauze oriented to give a vertical plane in the distillation column. The distillation packing pieces

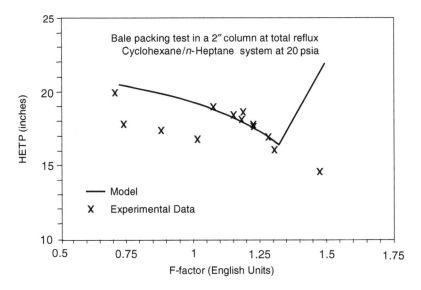

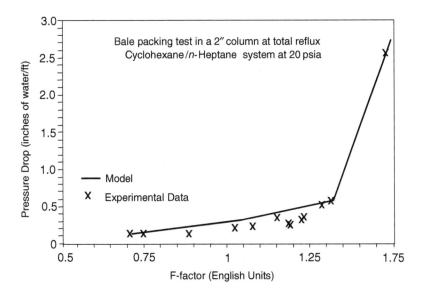

Figure 2.10: Mass transfer efficiency and capacity of Chemical Research & Licensing catalyst bales.

are about 1 foot (0.3 m) tall and installed as a monolith across the entire cross-section of the column. These pieces are stacked on top of one another to the desired height.

The structured reactive distillation packing (KATAMAX™ packing) consists of two pieces of rectangular crimped wire gauze sealed around the edge, thereby forming a pocket about 1–5 cm wide between the two screens. This pocket is filled with catalyst and the seal is completed, containing the catalyst between the screens. These screen *wafers* are bound together in cubes. The resulting cubes are transported to the distillation column and installed as a monolith inside the column to the required height.

Figure 2.11 shows a diagram of how the wafers are bundled into cubes and how the cubes are installed in the distillation column. When the catalyst is spent, the column is shut down and the packing is manually removed and replaced with packing containing fresh catalyst.

The advantages of this device are very efficient mass transfer between the liquid and vapor phases, low pressure drop, and high hydraulic capacity (DeGarmo et al., 1992). This allows for high throughput and fast mass transfer, giving the desired separation of products from reactants for equilibrium-limited reactions.

The regular, simple geometric design allows for accurate fundamental analysis of the mass transfer characteristics of the packing. Koch-Glitsch developed a rate-based mass transfer computer model specific to the packing (Pinjala et al., 1992). The model allows quick screening of potential new applications if the intrinsic kinetics and thermodynamics are known for the reaction.

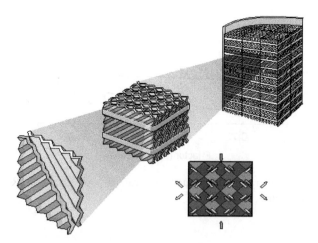

Figure 2.11:   Koch-Glitsch, Inc. KATAMAX packing installation.

As with the CR&L catalyst bales the KATAMAX packing can be loaded with almost any solid-formed catalyst and modified to different reactions depending upon the required loaded catalyst density. The wire gauze crimp in KATAMAX packing can be varied to allow more or less liquid hold-up.

KATAMAX packing has been applied to etherification for the production of MTBE, ethyl *tert*-butyl ether (ETBE), and tertiary amyl methyl ether (TAME) in the Huels Ethermax^{TM} process licensed by UOP LLC. Since 1992, seven licensed commercial units have been put into operation. The rate-based reactive distillation model was very successful in modeling the performance of these units (Frey et al., 1993).

2.6.3 Eastman Chemical High Liquid Hold-up Trays

Although this tray design is not in use outside of Eastman Chemical Company, it is worthy of mention as the most significant commercial application of homogeneous catalysis to reactive distillation. Eastman Chemical Company uses a specially designed high liquid hold-up tray to hold the homogeneous catalyst (sulfuric acid) that is used in the production of high-purity methyl acetate, as described in Section 2.2.1 (Agreda et al., 1990). The two-stage designs described above are suitable only for solid catalysts. The use of liquid sulfuric acid requires the use of bubble cap, reverse-flow trays with additional sumps to provide ample residence time of the reactants in the catalyst zone. With homogeneous catalysts greater care must be taken to keep the catalyst from escaping from the desired reaction zone and possibly entering an area of the column where undesirable reverse or side reaction may occur.

2.6.4 Other Leading Reactive Distillation Equipment Concepts

2.6.4.1 Trays or Downcomers Designed to Hold Solid Catalysts

Many inventors have proposed the use of a modified distillation tray and downcomer for reactive distillation (Carland, 1994; Jones, 1985; Quang et al., 1989; Nocca et al., 1989). The catalyst-containing tray concept has the advantages of high catalyst hold-up, well-established design criteria (although they must be modified for the inclusion of a solid phase), and intimate phase contacting. Additionally, trays could allow catalyst removal and replacement without emptying the column, as the catalyst could be kept fluidized and transported out of the column with the liquid

(Jones, 1995). Catalyst attrition is a concern in a fluidized environment, but this can be taken care of by filtration of the liquid and by make-up of the catalyst. Alternatively, downcomers filled with catalyst allow the solids to remain still as the liquid pours down a packed bed of catalyst. The primary drawback with installing the catalyst in downcomers is the limited volume available for catalyst inventory and the elimination of liquid–vapor mass transfer at the point of reaction.

The concept of incorporating a contained packed bed in a conventional distillation tray design has the benefits of using a device that has known mass transfer characteristics and is currently widely mass produced and has definite containment of the catalyst. The designs vary in the amount of distillation that occurs in the area of reaction.

At one extreme, nearly no liquid–vapor contacting is achieved in the reaction zone. This design has liquid pouring into a packed catalyst bed with the rising vapor essentially bypassed around the packed bed (Jones, 1992; Yang et al., 1994; Luebke et al., 1995). No fractionation is performed in the catalyst bed except for the vaporization caused by any heat of reaction. The cross-sectional area available for vapor flow determines the pressure drop. The primary experimental development required for such a device is the maintenance of good liquid flow distribution across the entire catalyst bed. A large packed bed could be made to have some vapor–liquid contacting features by containing the catalyst in miniature catalyst containers (Smith, 1984; Adams, 1991) or inserting void space into the packed bed (Buchholz et al., 1994).

Equipment has been suggested that could perform distillation and reaction within the same distillation column height in an otherwise conventional tray column. This usually consists of columns of catalyst contained by screens running through distillation trays vertically through the column (Pinaire et al., 1992; Yeoman et al., 1995, 1996). The difficulty in this design is to minimize the liquid that bypasses the distillation trays via the catalyst conduit between the trays while maintaining good catalyst–liquid contacting.

2.6.4.2 *Impregnating Catalyst on Conventional Distillation Equipment*

One of the more "talked about" methods for introducing catalyst into a distillation column has been to impregnate or coat a catalyst material on a common distillation tray or packing (Smith, 1981a). This concept has not been put into practice for the following reasons:

1. The amount of catalyst that can be loaded in a column in this manner is small compared to addition of catalyst pills or homogeneous catalyst.
2. Coating or impregnation of catalyst materials on metal surfaces is expensive.
3. Production of catalyst materials in the shape of distillation packings is also expensive.
4. There is currently no generic manufacturing method that can economically produce different catalyst materials as coatings or structured packings.

2.7 Conclusions

In the past 20 years, reactive distillation has emerged from being a subject of interest only to a few specialists to become one of the most successful new process technologies in the refining and bulk chemicals industries. Recent commercial applications in esterification and etherification have demonstrated that reactive distillation can be used for large production volumes and can cope with impure feedstocks. The technology that has been developed in these applications can now be applied to a range of other chemicals and processes.

The modeling and simulation of reactive distillation have also seen significant advances. Most commercial simulation programs now offer reactive distillation modules. Together with simple shortcut approaches, these simulation tools allow design engineers to evaluate the feasibility of reactive distillation in a fraction of the time that was previously needed. Theoretical understanding of the stationary points in reactive systems has also improved and continues to be the subject of academic research.

Until now, reactive distillation equipment has been designed for versatility to handle a variety of solid catalyst pills or granules. Production of a catalyst in a particular shape for bulk-phase mass transfer seems unlikely because of the development costs that would be required for each catalyst application. Existing solid catalysts can be easily accommodated without extensive modification by using catalyst container designs such as the Chemical Research & Licensing bales and the Koch-Glitsch KATAMAX packing. The design of catalyst containment devices will continue to be an area of ongoing development, with particular attention being paid to designs that allow for removal or in-situ regeneration of the catalyst.

Symbols

a	Interfacial area per unit volume
$C_{A,bulk}$	Concentration of A in the bulk liquid
C_{Ai}	Concentration of A at the interface
C_B	Concentration of B in the bulk liquid
D_{AG}	Vapor-phase diffusivity of A
D_{AL}	Liquid phase diffusivity of A
Da	Damköhler number, defined by equation 2.5
E	Enhancement factor for reactive absorption relative to physisorption
H	Henry's law constant
Ha	Hatta number, defined by equation 2.20
H_0	Liquid molar hold-up
k_1	Pseudo-first-order rate constant
k_g	Vapor-phase mass transfer coefficient
k_L	Liquid-phase mass transfer coefficient
K	Reaction equilibrium constant
N_{AG}	Flux of A through the vapor-phase boundary layer
N_{AL}	Flux of A through the liquid-phase boundary layer
p_A	Partial pressure of A in the bulk vapor
p_{Ai}	Partial pressure of A at the interface
r_A	Rate of reaction
V	Vapor molar flow rate
x_i	Mole fraction of component i

Greek

δ_G	Vapor-phase boundary layer thickness
δ_L	Liquid-phase boundary layer thickness

References

Adams, J.R., U.S. Patent 5,057,468 to Chemical Research & Licensing Company, 1991.

Agreda, V.H. and L.R. Partin, U.S. Patent 4,435,595 to Eastman Kodak, 1984.

Agreda, V.H., L.R. Partin, and W.H. Heise, *Chem. Eng. Prog.*, **86**(2), 40, 1990.

Arganbright, R.P., D. Hearn, E.M. Jones, and L.A. Smith, *Novel Process for Methyl Tertiary-Butyl Ether*, DOE/CS/40454-T3, 1986.

Backhaus, A.A., U.S. Patent 1,400,849, 1921a.

Backhaus, A.A., U.S. Patent 1,400,850, 1921b.

Backhaus, A.A., U.S. Patent 1,400,851, 1921c.

Backhaus, A.A., U.S. Patent 1,400,852, 1921d.

Backhaus, A.A., U.S. Patent 1,403,224, 1922a.

Backhaus, A.A., U.S. Patent 1,403,225, 1922b.

Backhaus, A.A., U.S. Patent 1,425,624, 1922c.

Backhaus, A.A., U.S. Patent 1,425,625, 1922d.

Backhaus, A.A., U.S. Patent 1,425,626, 1922e.

Backhaus, A.A., U.S. Patent 1,454,462, 1923a.

Backhaus, A.A., U.S. Patent 1,454,463, 1923b.

Barbosa, D.A.G. and M.F. Doherty, *Chem. Eng. Sci.*, **43**, 541, 1987.

Barchas, R., R. Samarth, and G. Gildert, *Fuel Reformulation*, **3**(5), 44, 1993.

Buchholz, M., R. Pinaire, and M.A. Ulowetz, U.S. Patent 5,275,790 to Koch Engineering Company, 1994.

Carland, R.J., U.S. Patent 5,308,451 to UOP, 1994.

Crossland, C.S., G.R. Gildert, and D. Hearn, U.S. Patent 5,431,890 to Chemical Research & Licensing Company, 1995.

DeGarmo, J.L., V.N. Parulekar, and V. Pinjala, *Chem. Eng. Prog.*, **88**(3), 43, 1992.

Doherty, M.F. and J.D. Perkins, *Chem. Eng. Sci.*, **33**, 281, 1979.

Frey, S.J., S.M. Ozmen, D.A. Hamm, V. Pinjala, and J.L. DeGarmo, "Advanced Ether Technology Commercialized," presented at the World Conference on Clean Fuels and Air Quality Control, Washington, D.C., October 6–8, 1993.

Hearn, D., G.R. Gildert, and E.M. Jones, Jr., U.S. Patent 5,523,062 to Chemical Research & Licensing Company, 1996.

Hickey, T.P. and J.R. Adams, U.S. Patent 5,321,163 to Chemical Research & Licensing Company, 1994.

Johnson, K.H., U.S. Patent 5,189,001 to Chemical Research & Licensing Company, 1993.

Jones, E.M., Jr., U.S. Patent 4,536,373 to Chemical Research & Licensing Company, 1985.

Jones, E.M., Jr., U.S. Patent 5,130,102 to Chemical Research & Licensing Company, 1992.

Jones, E.M., Jr., European Patent 0 402 019 B1 to Chemical Research & Licensing Company, 1995.

Lander, E.P., J.N. Hubbard, and L.A. Smith, Jr., *Chemical Eng.*, **90**(8), 36, 1983.

Luebke, C.P., T.L. Marker, and G.A. Funk, U.S. Patent 5,449,501 to UOP, 1995.

Nocca, J.L., J. Leonard, J.F. Gaillard, and P. Amigues, U.S. Patent 4,847,431 to Institut Francais Du Petrole, Elf France, 1989.

Pinaire, R., M.A. Ulowetz, T.P. Nace, and D.A. Furse, U.S. Patent 5,108,550 to Koch Engineering Company, 1992.

Pinjala, V. et al., "Rate-Based Modeling of Reactive Distillation Systems," presented at the AIChE 1992 Annual Meeting, Miami Beach, Florida, November 1–6, 1992.

Quang, D.V., P. Amigues, J.F. Gaillard, J. Leonard, and J.L. Nocca, U.S. Patent 4,847,430 to Institut Francais Du Petrole, Elf France, 1989.

Shoemaker, J.D. and E.M. Jones, Jr., *Hydrocarbon Processing*, **66**(6), 57, 1987.

Smith, L.A., U.S. Patent 4,250,052 to Chemical Research & Licensing Company, 1981a.

Smith, L.A., U.S. Patent 4,307,254 to Chemical Research & Licensing Company, 1981b.

Smith, L.A., Jr., U.S. Patent 4,443,559 to Chemical Research & Licensing Company, 1984.

Smith, L.A., Jr., U.S. Patent 5,120,403 to Chemical Research & Licensing Company, 1992.

Song, W., R.S. Huss, M.F. Doherty, and M.F. Malone, *Nature*, **388**, 561, 1997.

Subawalla, H., J.C. Gonzalez, A.F. Seibert, and J.R. Fair, *Ind. Eng. Chem. Res.*, **36**(9), 3821, 1997.

Ung, S. and M.F. Doherty, *AIChE J.*, **41**, 2383, 1995.

Venimadhavan, G., G. Buzad, M.F. Doherty, and M.F. Malone, *AIChE J.*, **40**(11), 1814, 1994.

Widagdo, S. and W.D. Seider, *AIChE J.*, **42**, 96, 1996.

Yang, Z., X. Hao, and J. Wang, U.S. Patent 5,308,592 to China Petrochemical Corporation (SINOPEC), 1994.

Yeoman, N., R. Pinaire, M.A. Ulowetz, T.P. Nace, and D.A. Furse, U.S. Patent 5,454,913 to Koch Engineering Company, 1995.

Yeoman, N., R. Pinaire, M.A. Ulowetz, T.P. Nace, and D.A. Furse, U.S. Patent 5,496,446 to Koch Engineering Company, 1996.

Chapter 3

EXTRACTION WITH REACTION

Vincent Van Brunt and Jeffrey S. Kanel

3.1 Introduction

Extraction is the unit operation that partitions at least one solute between two partially miscible liquid phases that differ in bulk density by approximately an order of magnitude or less. It is often coupled with reaction mechanisms to improve the separation efficiency of the basic unit operation. Both reversible and irreversible reactions are used. In practice, reactive extraction should be considered when: the reactants are relatively immiscible, the product undergoes subsequent undesired reaction in the reaction phase, the reaction products to be separated are immiscible with the reaction phase, the phase equilibrium can be positively influenced, the heat transfer is to be improved during the reaction, or the product–catalyst separation can be effected by a liquid–liquid separation. Industrial applications of reactive extraction are found in the petrochemical, pharmaceutical, metallurgical, nutraceutical, and nuclear industries. Hydrometallurgical extraction utilizes both chelations and ion-pair formation as active chemical mechanisms for extraction. Synergism and adduct formation are used to both improve selectivity and to augment solute loading. Dissociative extraction of organics also depends on the enhanced efficiency of reaction to provide viable separation methods for isomers.

Products that are produced by reactive extraction include: polyurethanes, polycarbonates, polyamides, soaps, linear alkylbenzene sulfonates, cephalosporin antibiotics, TNT, resorcinol, gasoline, fuel, and liquid petroleum gas.

The combination of extraction with a reaction comprises a large class of industrial processes. The strength of the operation relies on the unique combination of the partially miscible dense phases with at least one reaction. On one hand, the separation process is enhanced by reversible reactions to provide the preferred means of obtaining highly efficient hydrometallurgical mineral and metal purifications. On the other hand, superimposing a liquid–liquid equilibrium enhances liquid-phase reactions. The presence of a second dense fluid phase permits reactions to proceed beyond their single-phase limiting equilibrium by providing a way to have higher concentration reactants and by simultaneously providing a sink for one or more products. This technique is used in both inorganic and organic systems to increase the yield of a reaction. In addition, reactive extraction processes are the only efficient means to control highly exothermic reactions including nitrations, oxidations, and sulfonations.

In the case of metals purification, extraction is enhanced with specific reaction chemistries. In the case of many organic reacting systems, extraction is used to increase the efficiency of a liquid-phase reaction. In theory, limiting reactions can be employed, but in practice, reactions are usually assumed to be instantaneous. The second direct contact liquid phase provides a way to control a reaction. The second phase can provide a way to add incompatible reactants at greater than their single-phase solubility, and it can provide a sink for a product or byproduct permitting a reaction to progress to completion. The combined liquid phases provide a means to readily remove heat from exothermic reactions, which is not possible with a vapor–liquid separation.

3.2 Hydrometallurgical Separations

It is unlikely that an electrolyte, cation, or anion will leave an aqueous phase and enter a nonpolar organic phase unless some additional driving force is present. The reaction of metals or other cations and anions to form organic complexes enables them to be transported from the aqueous phase to the organic phase. The design of chemical complexation reactions for specific metals, cations, or anions enables them to be separated from a mixture of other similar species. This is the basis of viable processes for the recovery of metals from ore leachates. These processes rely on reversible chemical reactions to both recover the metal and to regenerate the solvent. Broad overviews covering the breadth of the chemistries for specific metals have been written by Marcus and Kertes (1969) and by Sekine and Hasegawa (1977).

The two-volume text by Ritcey and Ashbrook (1984) describes both fundamentals and specific applications.

Hydrometallurgical extractions rely on a solvent composed of a *diluent* that contributes the bulk physical properties and at least one chemical capable of reversibly complexing with a solute known as an *extractant*. The reaction between the solute, in the case of a metal, usually an anion or cation, and the extractant is usually quantitative. Table 3.1 shows some of the alternative types of extractants in use. Multiple extractants can sometimes have profound effects that include the formation of soluble adducts that radically affect the equilibrium between the solute and the solvent. This phenomenon, known as *synergism*, can increase an equilibrium separation by orders of magnitude (Blumberg, 1988; Lo et al., 1983; Marcus and Kertes, 1969; Ritcey and Ashbrook, 1984).

These concepts are shown in a typical solvent extraction cycle in Figure 3.1. In this example zirconium is separated from hafnium by using a solvent that includes tributyl phosphate (TBP) in a hydrocarbon diluent such as dodecane. The feed from an acid ore leachate includes both +4 cations in aqueous nitric acid and sodium nitrate. The feed is separated into a *raffinate* containing almost all the hafnium and an *extract* that contains almost all the

Table 3.1: Types of Extraction Solvents Typically Used for Reactive Extraction Processes.

Type/Example	Reaction Mechanism	Application
Phosphoric, phosphonic, phosphinic acids/di(2-ethylhexyl)phosphoric acid/(D2EHPA)	Complexation (acidic)	Co/Ni, U, rare earths
Carboxylic acids/naphthenic,versatic	Complexation (acidic)	Cu/Ni, Co, Fe, Zn
Sulfonic acids/dinonylnaphthalene sulfonic acid (DNNSA)	Complexation (acidic)	Mg
Primary amines/1,1,3,3,5,5,7,7,9,9-decamethyl decamine	Ion-pair formation (basic)	Th, U
Secondary amines/ditridecylamine	Ion-pair formation (basic)	U
Tertiary amines/trioctylamine (TOA), triisooctylamine (TIOA)	Ion-pair formation (basic)	U, W, V
Quaternary ammonium salts/ trioctylmethylammonium chloride	Ion-pair formation (basic)	Mo, W, V, U

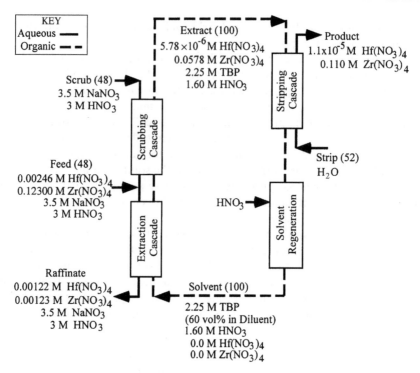

Figure 3.1: Zirconium/hafnium separation utilizing extraction, stripping, and scrubbing sections. Numbers in parentheses indicate relative mass flow rates. (Adapted from Dorf, 1996.)

zirconium. In the extraction cascade, most of the zirconium and some of the hafnium is extracted into the solvent by complexing with the TBP. In the scrubbing cascade, the hafnium is preferentially back extracted from the solvent into the aqueous phase. The scrubbing cascade is analogous to a rectification section of a distillation column, and each stage in the scrubbing section further purifies the extract by reducing the level of the coextracted hafnium. The scrub solution is analogous to a distillation column's reflux; however, in addition to being able to control its temperature and flow rate, the scrub solution chemistry can be controlled. Both its pH and its anion concentration can be controlled independently. The nonextractable sodium nitrate, known as a *salting-out agent*, provides a source of nitrate ion, permitting its concentration to be greater than that from the nitric acid alone. The capacity of the extract, called its *loading*, is the ratio of the solute concentration to the extractant concentration. The solutes are unloaded from the solvent in the stripping cascade by contacting the solvent with a

strip solution (in this case water). The solvent is then returned to the extraction cascade after being periodically regenerated.

This basic cycle is used in most mineral and metals purifications repeatedly. It may be modified as needed by adjusting the relative flows, temperatures, and chemistries in each of the cascades.

The concept of multistage extraction is illustrated in Figure 3.2. In this
flowsheet, the 1A scrub can be at a different temperature and use a different
chemistry than the 2A scrub. In combination, the sequential scrubbing cascades can affect a relatively pure first-cycle extract product. This product is
sent to stripping. In the stripping cascade A, the solute is recovered in the
first-cycle product. This is sent to a solvent scrubbing cascade in which the
solvent phase is used to further purify the strip. Between the two scrubbing
cascades and the solvent scrubbing, the first-cycle product is obtained. A
higher concentration can be obtained by cascading these cycles. In this
process, the first-cycle (A) product is sent to a second cycle (B) in which
the extraction–scrubbing–stripping cascades are repeated. Up to four cycles
have been used. In each cycle the separation factor between the two solutes
is amplified. Note that the second-cycle raffinate will usually have solute
concentrations that are higher than the first-cycle raffinate; hence, it is prac-

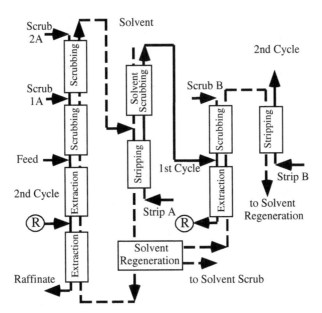

Figure 3.2: Two-cycle extraction process used for mineral and metals purification
with solvent scrubbing of the first-cycle product.

tical to recover them by returning the second-cycle raffinate to the appropriate point in the first-cycle extraction cascade. Although this concept is illustrated with this single stream, the same concept can be applied to other process streams as well.

The PUREX^{TM} process (Swanson, 1991; McKay et al., 1991; Frantz and Van Brunt, 1987; O'Quinn and Van Brunt, 1987) uses the difference between the extractability of +4 and +3 cations in TBP (Schultz and Navratil, 1984–1991) to separate plutonium from uranium after they are extracted from nuclear waste. The waste products have +1, +2, or +3 charges. Plutonium can exist in the +3 or +4 state, and uranium can be in the +4 or higher oxidation state. The process, shown in Figure 3.3(a), first extracts the +4 plutonium and the uranium from the feed, since TBP will extract the +4 and higher oxidation state metals. Second, the scrubbing stages exploit an irreversible aqueous phase reaction to reduce the plutonium to the nonextractable +3 state. This permits a

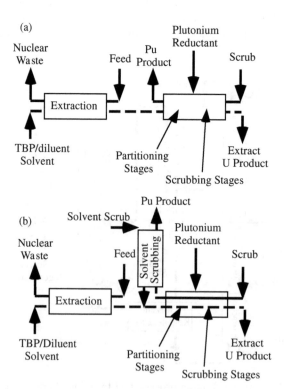

Figure 3.3: (a) Partitioning of U and Pu. (b) Solvent scrubbing of Pu product.

separation of the plutonium from the uranium in a process known as partitioning.

The irreversible reduction reaction of the plutonium is superimposed on an extraction and scrubbing process. Using the principle previously introduced, the aqueous plutonium product can be further purified by coupling a solvent scrubbing cascade to this flowsheet, see Figure 3.3(b). In this case, the partitioning stages can be considered to both scrub the uranium product and to react and strip the plutonium product from the solvent.

The PUREX^{TM} process illustrates the useful combination of irreversible reactions with reversible reactions. It also illustrates how fractional extraction principles can be eloquently coupled to reactive extraction flowsheets.

Hydrometallurgical separations rely on both the aqueous chemistry and organic phase reactions to enhance the selectivity and the solute capacity of a solvent. This is also the basis for many organic reacting systems that rely on chemical complexation to increase product recovery, e.g., penicillin G. In some organic systems a separation can be effected by exploiting the dissociation in an aqueous phase of one species over another, e.g., *m*-cresol dissociates in caustic more readily than *p*-cresol. In this process known as dissociative extraction, the reversible dissociation reactions are the basis of the isomer separation (Hanson et al., 1974). In many predominantly organic systems, the primary consideration is usually a reaction that is in some way limited in a single liquid phase. Often, the second phase provides a catalyst that is recycled. Occasionally, the product is formed in a second phase that is easily removed from the reacting phase.

3.3 Homogeneous Reactions: Organic Separations

3.3.1 Introduction

This section describes a sampling of industrial organic reactions that are carried out in two liquid phases. In general, the subject comprises both inorganic and organic reactions that are dominated by one or more irreversible reactions. The inorganic irreversible reactions include the production of salts such as potassium nitrate from potassium chloride and nitric acid (Blumberg, 1988; Hanson, 1971). Several generalized studies have been published on methods to identify the reaction mechanism and the extent of reaction (Piret et al., 1960; Trambouze and Piret, 1961; Trambouze et al., 1961; Sharma, 1991; Doraiswamy and Sharma, 1984). Theoretically, the reaction zone can be identified, as can the relative rates of reaction and mass transfer. In practice the processes are complex. Reaction kinetics and mass transfer from and to drops must be superimposed on gross drop movement while considering drop size distribution as well. The focus here is on major classes of reactive extraction in use at this time. Sulfonation and other historical examples are considered elsewhere (Doraiswamy and Sharma, 1984; Knaggs and Nepras, 1982).

3.3.2 Hydrolysis

Hydrolysis of fats (esters of fatty acids with glycerol, or triglycerides) and waxes (esters of fatty acids with fatty alcohols) under liquid–liquid conditions is used industrially to produce fatty acids (Brockmann et al., 1987). Vegetable fats from seeds are a common raw material for this production, although fruit and animal fats are also used. Fat splitting may be accomplished with water (hydrolysis), methanol (methanolysis), caustic soda (saponification), and amines (aminolysis). Soaps and fatty acid amides are predominately produced from fatty acids and fatty acid methyl esters (Brockmann et al., 1987, p. 254). Other commercial applications of hydrolysis include the manufacture of benzyl alcohol by the alkaline hydrolysis of benzyl chloride, see Doraiswamy and Sharma (1984, p. 50). An unwanted heterogeneous hydrolysis reaction occurs when acetic acid is extracted from an aqueous stream using an acetate solvent. The acetate hydrolyzes to produce an alcohol, which may be removed by the method described by Jones and Fallon (1997). Biomass, such as cellulose acetates, may also be subjected to hydrolysis in a liquid–liquid environment, according to Converse and Grethlein (1985). Hydrolysis reactions that are discussed in the literature are summarized in Table 3.2.

Table 3.2: Summary of the Literature for Liquid–Liquid Hydrolysis Reactions.

Reactants	Catalysts	References
2-Chloro-2-phenylacetic acid	NaOH and PTC	Starks et al. (1994)
Methyl acetate, ethyl acetate, isopropyl acetate	None	Jones and Fallon (1997)
Butyl acetate	NaOH	Sarkar et al. (1980); Sharma and Nanda (1968); Alwan et al. (1983)
n-Amyl acetate	None	Alwan et al. (1983)
iso-Amyl acetate	None	Alwan et al. (1983)

Table 3.2 (*continued*)

Reactants	Catalysts	References
n-Amyl acetate, *n*-hexyl acetate	NaOH	Hiraoka et al. (1990)
Vinyl dibromides	Carbon tetrabromide, TPP, NaOH, PTC-Pd(diphos)$_2$	Starks et al. (1994)
Ethyl chloroacetate	NaOH	Nanda and Sharma (1967)
Ethyl dichloroacetate	NaOH	Nanda and Sharma (1967); Sharma and Nanda (1968)
Methyl dichloroacetate	NaOH	Sarkar et al. (1980); Sharma and Nanda (1968); Alwan et al. (1983)
Methyl trichloroacetate	NaOH	Nanda and Sharma (1967)
Butyl formate	NaOH	Sarkar et al. (1980); Bhave and Sharma (1981)
iso-Amyl formate	NaOH	Nanda and Sharma (1967); Sharma and Nanda (1968)
2-Ethyl hexyl formate	None	Bhave and Sharma (1981)
n-Octyl formate	None	Bhave and Sharma (1981)
iso-Decyl formate	NaOH, or NaOH + Na$_2$SO$_4$	Bhave and Sharma (1981)
Dodecyl formate	NaOH + Na$_2$SO$_4$	Bhave and Sharma (1981)
Mixed triglycerides (palmitic, stearic, oleic, beeswax, etc.)	NaOH	Jeffreys et al. (1961)
Polyfluorobenzene	NaOH + tetrabutyl ammonium salt	Starks et al. (1994, p. 368)
Haloaromatic ketones	None	Starks et al. (1994, p. 368)
Partially crosslinked copolymer of methyl methacrylate with divinylbenzene	18-Crown-6, [2.2.2] cryptand + PEG	Starks et al. (1994, p. 495)

In hydrolysis, water is reacted with another compound (such as an organic acid or ester) to form two or more new substances. Hydrolysis reactions may be catalyzed under either alkaline or acidic conditions (Morrison and Boyd, 1973, p. 663); the overall stoichiometry of the alkaline hydrolysis of an ester is

$$RCOOR' + OH^- \leftrightarrow RCOO^- + R'OH \tag{3.1}$$

The ester and hydroxyl components are typically contained in the nonpolar and polar phases, respectively. The phase which the products favor depends on the nature of the R or R$'$ groups. The hydrolysis of a triglyceride occurs by the following equation, where the R groups may be different:

$$
\begin{array}{ll}
CH_2COOR & CH_2OH \\
| & | \\
CHCOOR + 3H_2O \rightarrow & CHOH + 3RCOOH \\
| & | \\
CH_2COOR & CH_2OH
\end{array}
\tag{3.2}
$$

Industrial fat hydrolysis involves a catalyst that may be an acid, lipases, or basic oxides. Continuous splitting processes operate at 210 to 260°C and 1.9 to 6 MPa (Brockmann et al., 1987, p. 258). Badger, Foster-Wheeler, and Lurgi have specially designed continuous countercurrent splitting towers.

Doraiswamy and Sharma (1984, p. 50) stated that benzyl alcohol was manufactured by alkaline hydrolysis of benzyl chloride via the reaction $C_6H_5CH_2Cl$ (organic) + OH^- (aqueous) $\rightarrow C_6H_5CH_2OH + Cl^-$. The reaction rate was very slow, so the diffusional mass transfer factors were not of importance. Buysch et al. (1998) claimed a process for the production of benzyl alcohol by hydrolysis of benzyl chloride with water between 80 and 180°C, and with a conversion of 30 to 90%. Hag and Rantala (1984) claimed a process for the continuous hydrolysis of an α-chlorinated toluene compound contained in the organic phase that is passing countercurrently to an aqueous phase containing the hydrolyzing agent.

The synthesis of the β-lactam antibiotic, imipenem, utilizes extractive hydrolysis for one stage of its synthesis, according to Verrall (1992, pp. 237–239). The conversion of the *o*-formyl group to the corresponding alcohol was originally performed with methanol as a miscibilizing agent. However, the solvent recovery costs were unacceptable, and it was noticed that a majority of the impurities were insoluble in water. Therefore, the separation of the product from impurites without the use of an additional solvent was effected by forming the water-soluble enolate salt. The aqueous alkali hydrolysis was very rapid, but both the product and the reactant were unstable in the alkaline solution. Therefore, a centrifugal extractor was employed to minimize the residence time of the dispersion in the extractive hydrolysis step of the process. Details of the extractive hydrolysis are shown in Figure 3.4, where Compound I is found in conjunction with dichloromethane, triphenylphosphine oxide, the hydrazide, and formic acid. Compound I is fed with dichloromethane, countercurrently, to an aqueous caustic solution. Compound II is produced and extracted into the aqueous phase.

Dehmlow and Dehmlow (1993, pp. 215–218) summarized a significant number of hydrolysis reactions including: (1) the conversion of propylene oxide to 1,2-propanediol nearly free of polyglycols when heated with water and methyltributylammonium iodide under carbon dioxide (p. 215); (2) the hydrolysis of sulfuryl chlorides in a liquid–liquid system with quaternary ammonium salts; (p. 217); and (3) the partial hydrolysis of nitriles to amides in pyridine/water/potassium hydroxide using NBu_4Br or in dichloromethane/water/sodium hydroxide/30% H_2O_2 using NBu_4HSO_4 (p. 217).

A variety of hydrolysis reactions that utilize phase transfer catalyst (PTC) were referenced by Starks et al. (1994): (1) hydrolysis of beeswax with an aqueous caustic solution using polyethylene glycol (PEG) as the phase transfer catalyst (p. 166); (2) alkaline hydrolysis of a sterically hindered polymeric ester, such as partially crosslinked polymethyl methacrylate, using 18-crown-6, the [2.2.2] cryptand, and PEG as the phase transfer catalyst (p. 495); and (3) vinylic dibromide hydrolysis to carboxylic acids by PTC-Pd(diphos)$_2$ (p. 617). The latter provided a route for a one-carbon

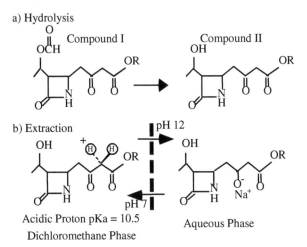

Figure 3.4: Extraction of imipenem. Left side depicts hydrolysis of formyl intermediate imipenem. Right side illustrates the extraction of the imipenem intermediate of the enolate salt. (From Verrall, 1992, p. 238.)

addition to a carbonyl group, ending with a carboxyl group
($C_6H_5CHO \rightarrow C_6H_5CH_2COOH$, for example).

3.3.3 Saponification

Saponification occurs when an ester is heated with aqueous alkali to form an
alcohol and the salt of the acid corresponding to the ester. Industrial applications of saponification include the production of soap, as described by
Jeffreys et al. (1961). A number of studies on saponification are presented in
Table 3.3. Generally, the overall stoichiometry of saponification is

$$RCOOR' + NaOH \leftrightarrow RCOO^-Na^+ + R'OH \tag{3.3}$$

Jeffreys et al. (1961) investigated the performance of a commercial-scale
continuous countercurrent fat splitting (Treybal, 1963, p. 570) by computing
the height equivalent to a theoretical plate (HETP) and the height of a
transfer unit (HTU). Traditionally, soap production was performed batchwise by saponification of animal fat with caustic alkali to obtain soap and

Table 3.3: Summary of Saponification Reactions Performed as Reactive
Extractions.

Saponification of:	Catalyst	Reference
n-Amyl acetate	None	Hiraoka et al. (1990)
n-Hexyl acetate	None	Hiraoka et al. (1990)
Diacetoxymethylbenzene	None	Maeda et al. (1976)
Methyl esters	None	Gupta (1995)
Dimethyl adipate	Onium salt (Aliquat 336)	Dehmlow and Dehmlow (1993, p. 219)
Diethyl adipate	Aliquat 336	Dehmlow and Dehmlow (1993, p. 219)
Various esters	Triphase catalyst	
Methyl and octyl mesitoate	Aliquat 336	
Complex *tert*-butyl ester	NBu$_4$HSO$_4$	
Diethyl dialkylmalonate	NR$_4$X	Dehmlow and Dehmlow (1993, p. 221)
β-Keto esters	H$_{33}$C$_{16}$NMe$_3$Br	
Fats and oils	Aliquat 336	
Waxes	Dibenzo-18-crown-6	
Heteroaryl acetonitrile	NBu$_4$X	
Dihydrovitamin K$_1$ diester	None	Dorner (1998)

glycerin, and free fatty acids were obtained by acidifying the soap. According to Jeffreys et al., continuous soap manufacture consisted of feeding water droplets (450–500°F and 600–700 psi) countercurrently to the fats. A small amount of water was dissolved in the fat phase and reacted as follows:

$$
\begin{array}{l}
CH_2COOR \\
| \\
CHCOOR \ + 3H_2O \\
| \\
CH_2COOR
\end{array}
\rightarrow
\begin{array}{l}
CH_2OH \\
| \\
CHOH \ + 3RCOOH \\
| \\
CH_2OH
\end{array}
\tag{3.4}
$$

The resulting glycerin distributed preferentially into the aqueous phase. Hydrolysis was between 97 and 99.5% complete in the column. Therefore, the continuous phase leaving the top of the extraction column was predominately free fatty acids with only trace impurities that were removed by distillation. The distillate was neutralized with soda ash, and finished soap was produced by pH adjustment with caustic soda. Jeffreys et al. found that the values for HETP and HTU with reaction were much greater than would be anticipated in the absence of reaction. Concentration profiles of glycerin and fat in the column were predicted from the developed models. They recommended the use of a mechanically agitated column instead of a spray column to reduce the column height. Their method for computing HTU and HETP is believed to be applicable to many other reactive extraction processes. Brockmann et al. (1987, p. 254) note that ricinoleic acid and hydroxystearic acid are commercially produced by the saponification of castor oil and hardened castor oil, respectively.

Dehmlow and Dehmlow (1993, pp. 218–222) noted that crown ethers and cryptates were valuable in the saponification of sterically hindered esters that were difficult to hydrolyze under standard conditions. For the saponification of diethyl adipate with 50% aqueous sodium hydroxide/Aliquat 336, the more nonpolar solvent was found to be more useful. Varying the catalyst anion for the saponification of diethyl adipate in petroleum ether, the effectiveness increased in the series: I < Br < Cl < hydrogen sulfate < hydrogen adipate. The investigation of symmetrical cations on yield for the standard conditions of this reaction showed the following yields: 18, 20, 45, 46, and 39%, for $N(C_4H_9)_4Br$, $N(C_5H_{11})_4Br$, $N(C_6H_{13})_4Br$, $N(C_7H_{15})_4Br$, and $N(C_8H_{17})_4Br$, respectively.

Dorner (1998) claimed a process for the production of vitamin K_1, wherein dihydrovitamin K_1 acetate benzoate was saponified with sodium hydroxide in a two-phase dispersion. The polar solvent comprised aqueous methanol and sodium hydroxide and the nonpolar phase comprised petroleum ether and the dihydrovitamin K_1. After several subsequent processing steps, the vitamin K_1 was separated from the nonpolar phase via filtration.

3.3.4 Esterification

Reacting a carboxylic acid and an alcohol with the elimination of water produces esters. Since the esterification of an organic acid and an alcohol involves a reversible equilibrium, the removal of either product can drive the conversion close to 100% if the reaction is equilibrium limited instead of rate limited. The overall stoichiometry of acid catalyzed esterification reactions was reported by Morrison and Boyd (1973, p. 520) as

$$ROH + R'COOH \rightarrow R'COOR + H_2O \tag{3.5}$$

Cephalosporin antibiotics are produced through the intermediate cephalosporin C, which is manufactured via fermentation and is amphoteric in contrast to penicillin G, which is acidic (Verrall, 1992, pp. 234–237; Ridgway and Thorpe, 1991, pp. 589–590). The acidity of the penicillin allows it to be transferred from an aqueous phase into a nonpolar phase by liquid extraction. For cephalosporin C, blocking the 5′-amino group with a suitable component permits it to be extracted into a nonpolar phase. However, the acetoxy group on the 3-side chain (see Figure 3.5) may undergo hydrolysis to the 3-hydroxymethyl compound, which forms the lactone under acetic conditions. To prevent lactone formation during extraction, extractive esterification has been utilized. The amino group is blocked with an acid chloride or chloroformate ester (preferred due to its lower hydrolysis rate in an aqueous medium) at 0 to 5°C and pH 8. The aqueous solution, which contains the protected cephalosporin C, is intimately contacted with a dichloromethane phase containing a substituted diazomethane (such as diphenyldiazo-

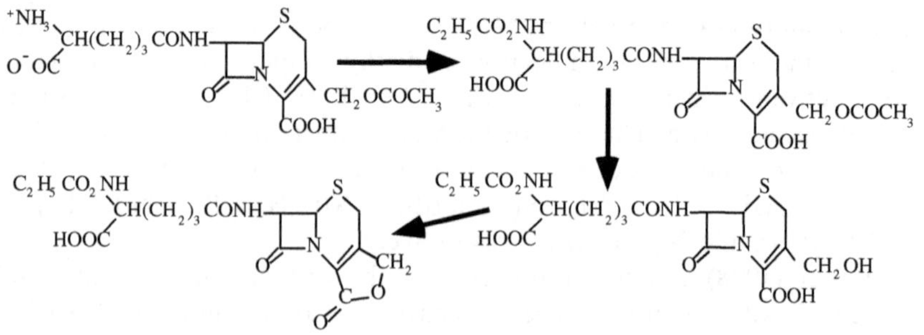

Figure 3.5: Structure of cephalosporin C and the process under which it forms the lactone (Verrall, 1992).

methane). The aqueous phase is acidified to between pH 2 and 2.5, and the bis(diphenylmethyl)ester is transferred into the nonpolar phase at 20°C. The required residence time for the extractive esterification is approximately one hour. Choice of the extractive–esterifying agent, the diazoalkane, is primarily constrained by cost and stability.

Interfacial esterification of sucrose with benzoyl chloride to make sucrose benzoate, termed the Schotten–Baumann technique, is commercially practiced, according to Doraiswamy and Sharma (1984, p. 96). The aqueous sucrose solution is contacted with benzoyl chloride dissolved in toluene, and sodium hydroxide is added to neutralize the hydrochloric acid formed by the reaction.

Martin and Krchma (1930) patented a process for the conversion of acetic acid and butanol into butyl acetate involving a liquid–liquid system. Al-Saadi and Jeffreys (1981a–c) measured the phase equilibrium for the water/butanol/butyl acetate/acetic acid/sulfuric acid (catalyst)/heptane system. First, they measured the ternary systems and used Hand's method to correlate the data (assuming the solvent and the carrier fluid were totally immiscible). Then, the quaternary phase equilibrium was measured for the water/acetic acid/butanol/heptane system and the water/acetic acid (or butanol)/butyl acetate/heptane system. Correlations for both quaternary systems were developed, and were found to fit the experimental data well. The esterification kinetics were measured at 25°C for *n*-butanol with acetic acid in the presence of sulfuric acid and *n*-heptane (the organic-phase solvent). Rate data for the homogeneous reactions were applied to evaluate the performance of the two-phase batch reaction results. The effects of mass transfer and esterification rate were considered to develop a model for the reactive extraction. The extractive esterification was then performed in a pilot-scale rotating disc contactor that was 10.1 cm in diameter and 92 cm in length. The heavy phase that entered the top of the extractor comprised the water, acetic acid, and sulfuric acid. The light phase, which consisted of heptane and butanol, was charged to the base of the extractor. Unusual phase inversion behavior was observed in the countercurrent extractor. In the top section of the extractor, the aqueous phase, containing the acetic and sulfuric acids, was observed to be continuous. However, in the bottom section of the extractor, the continuous nonpolar heptane phase contained butanol. Both at the top and bottom of the column, the dispersion was stabilized by solute transfer from the continuous phase, acetic acid and butanol in the top and bottom sections, respectively. This stabilization phenomenon resulted from interfacial tension gradients that were described by Groothuis and Zuiderweg (1960), Davies and Jeffreys (1971), Smith et al. (1963), and Lawson and Jeffreys (1965). The continuous esterification of *n*-butanol and acetic acid with a sulfuric acid catalyst, and water and heptane as the solvent system was experimentally demonstrated and the results were predicted by a computer model.

Minotti (1998) developed a generalized model for reaction coupled with extraction that solved a set of model equations for the composition profile in a cascade of isothermal stirred tank reactor–extractors, which are operating in either cocurrent or countercurrent flow. The following parameters were investigated: number of reactor–extractor stages, solvent flow rate, saturating the solvent with reactants, changing the extent of reaction distribution along the cascade, and reactive-phase hold-up. Their model was evaluated for the production of butyl acetate by the esterification of acetic acid and butanol. The reaction occurred in the aqueous phase, and the butyl acetate was extracted into the butanol phase. It was determined that a solvent in a reactive extraction system should preferentially extract the products and not the reactants to achieve good performance.

3.3.5   Transesterification

Transesterification involves the reaction between an ester and a second compound in which alkoxy groups or acyl groups are exchanged to a different ester. Three types of transesterification reactions were identified by Aslam et al. (1994, p. 775):

- *alcoholysis*—exchange of alcohol groups as

$$RCOOR' + R''OH \rightarrow RCOOR'' + R'OH$$

- *acidolysis*—interchange of acid groups as

$$RCOOR' + R''COOH \rightarrow RCOOH + R''COOR'$$

- *ester–ester interchange*—where two esters react as

$$RCOOR' + R''COOR''' \rightarrow RCOOR''' + R''COOR'$$

Koyama et al. (1996) disclosed a process for the production of sucrose fatty-acid esters by reacting sucrose with a fatty-acid alkyl ester in the presence of an alkali catalyst by using dimethyl sulfoxide as a reaction solvent. The steps used were: (1) subjecting the reaction mixture to a first liquid–liquid extraction by using a barely water-miscible organic solvent (alcohol or ketone having at least four carbon atoms) and water as extractants, and regulating the pH value of the aqueous phase between 3 and 7.5 to extract dimethyl sulfoxide in the reaction mixture into the aqueous phase, and the sucrose fatty-acid ester into the organic solvent phase; (2) subjecting the organic solvent solution containing the sucrose fatty-acid ester obtained as the organic solvent phase in the first liquid–liquid extraction to a second liquid–liquid extraction by a continuous countercurrent system using water, to thereby extract the dimethyl sulfoxide remaining in the organic solvent solution into the aqueous phase, thus giving an organic solvent solution which is substantially free from dimethyl sulfoxide; and (3) recovering the sucrose fatty-acid ester from the organic solvent solution.

3.3.6 Polycarbonates

Bisphenol A polycarbonate (BPA-PC) was disclosed by Schnell et al. (1962) at Bayer and independently by Fox (1964) at General Electric. The polymer was first commercially produced during 1958 by Bayer in Germany and during 1960 by General Electric in the United States. After litigation, U.S. patents were issued to Bayer for the interfacial process for the preparation of polycarbonates. The transesterification process and polycarbonate products produced by that process were claimed in the basic General Electric patent. Globally, more than one million tons of polycarbonate are produced annually, and most BPA-PC is produced via the interfacial polymerization process, according to Brunelle (1996, p. 585). This process involves the reaction of bisphenol A with phosgene in a liquid–liquid dispersion (Serini, 1992, pp. 210–211).

$$(3.6)$$

or

$$(3.7)$$

The bisphenol A is dissolved in the aqueous phase as sodium bisphenolate. Phosgene is dissolved in the nonpolar phase, which generally comprises chlorinated hydrocarbons such as methylene chloride, chlorobenzene, or dichloromethane. A 10–20 mol% excess of phosgene is typically used. The reaction occurs at the liquid–liquid interface, where carbonate oligomers are formed and transfer into the nonpolar phase. The hydrolysis products, NaCl and Na_2CO_3, diffuse into the aqueous phase. After the carbonate oligomers are formed, a catalyst, such as a tertiary aliphatic amine, is added to accelerate the polycondensation of the oligomers to product and catalyze the hydrolysis of excess terminal chlorocarbonate groups that were formed due to the excess of phosgene. The pH of the aqueous phase during the polycondensation reaction is maintained at 9 to 14. The reaction may be performed batchwise or continuously in agitated vessels or tubular reactors. The aqueous alkaline phase, which contains sodium chloride and sodium carbonate, is decanted from the organic phase via centrifugation. The organic phase is washed with dilute aqueous sodium hydroxide and then with dilute aqueous acid (hydrochloric or phosphoric) to neutralize the alkali. Finally, the organic phase is washed with demineralized water. The polycarbonate is typically separated from the solvent in one of three ways: precipitation with a nonsolvent, such as heptane; precipitation with hot water and evaporation of the polycarbonate solvent; and evaporation of the solvent in an extruder. The catalyst for the polycondensation reaction may include hindered secondary amines, tertiary amines, quaternary amines, and their salts. The catalysts are typically used at 1–2 wt% with respect to the bisphenol A.

3.3.7 Nitration

Nitration of aromatic compounds is an industrially important example of a liquid–liquid reaction where one reactant is initially in each immiscible phase and the two products are separated into the two phases. Nitration results in the substitution of a $-NO_2$ group for a hydrogen atom in a C—H, C—OH, or N—H group, which produces an ester, or nitramine in the latter two cases, respectively. These reactions were summarized by Albright (1996, p. 68) as

$$C-H + HNO_3 \rightarrow C-NO_2 + H_2O \qquad (3.8)$$

$$C-OH + HNO_3 \rightarrow C-O-NO_2 + H_2O \qquad (3.9)$$

$$N-H + HNO_3 \rightarrow N-NO_2 + H_2O \qquad (3.10)$$

Free-radical reactions are used to nitrate paraffins, cycloparaffins, and olefins, while ionic nitrations are commonly used for aromatics, heterocyclics, amines, and hydroxyl compounds such as alcohols, glycols, glycerol, and cellulose. Ionic nitrations are typically performed in liquid–liquid dispersions.

Booth (1991, p. 412) summarized a host of industrially important compounds produced by nitrating benzene, toluene, chlorobenzene, xylenes, naphthalenes, phenol, and resorcinol. Only benzene, toluene, and chlorobenzene nitrations are performed industrially at the largest scale, and key intermediates derived from these compounds were identified by Booth (1991, pp. 416–417, 419, and 426, respectively). Nearly 95% of the nitrobenzene produced is converted to aniline, which has hundreds of downstream applications, including methylene diphenylene diisocyanate (MDI) and poly(methylene diphenylene isocyanate) (PMDI). These two compounds accounted for about 75% of the aniline demand in 1996 with a volume of about one billion pounds per year. Mono- and dinitrotoluenes are typically hydrogenated to the associated amines. Other nitrotoluenes and their applications include: 2-nitrotoluene, which is converted to colorant intermediates; 3-nitrotoluene, which is transformed into *m*-toluidine, an important coupling agent for azo dyes; 4-nitrotoluene derivatives, which are used for colorants; and 2,4-dinitrotoluene, which is hydrogenated and subsequently converted to toluene diisocyanate (TDI) for use in polyurethanes. The 2,4-dinitrotoluene may undergo further nitration for the production of trinitrotoluene (TNT). Applications for selected chloronitrobenzenes include: 2-chloronitrobenzene derivatives are used in the synthesis of colorants; 4-chloronitrobenzene is a precursor for a drug used to treat leprosy; 1-chloro-2,4-dinitrobenzene is used to produce insecticides and dyes; and 1,4-dichloro-2-nitrobenzene is used to produce colorant intermediates. Other nitrated aromatic compounds include xylenes, naphthalenes, phenol, and resorcinol. Nitrated xylenes produce nitroxylenes, which are reduced to the corresponding xylidines, which are used as starting materials for the production of riboflavin and agrochemicals, respectively, according to Booth (1991, p. 423).

Most aromatic nitrations are performed by intimately contacting an aqueous phase, which contains the acid catalyst, and an organic phase, which contains the compound to be nitrated. A mixture of nitric and sulfuric acids is typically used for the catalyst. Doraiswamy and Sharma (1984, p. 57), and Booth (1991, p. 413) note that the typical nitrating agent for large-scale aromatic mononitrations consists of 20% nitric acid, 60% sulfuric acid, and 20% water. The effect of the material ratio of these three components on reactor performance was summarized by Albright (1996, p. 70). Centrifugal separators are used in many modern processes to separate the acid and hydrocarbon phases to minimize the amount of nitrated products in the plant at a given time.

The NO_2^+ mechanism for nitration has been accepted since about 1950 for many aromatic hydrocarbons, glycerol, glycols, and other compounds, according to Albright (1996, p. 68). When sulfuric acid is used as a co-catalyst with nitric acid, the following ionization reactions occur:

$$2H_2SO_4 + HNO_3 \leftrightarrow NO_2^+ + HSO_4^- + H_2O \tag{3.11}$$

$$H_2SO_4 + H_2O \leftrightarrow HSO_4^- + H_3O^+ \tag{3.12}$$

$$2HNO_3 \leftrightarrow NO_2^+ + NO_3^- + H_2O \tag{3.13}$$

$$HNO_3 + H_2O \leftrightarrow NO_3^- + H_3O^+ \tag{3.14}$$

Equilibrium constants were reported for the ionization reactions by Albright (1996, p. 69). The NO_2^+ attacks the aromatic compound (ArH) as shown below, where the aromatic (ArH) reacts with the NO_2^+ that is in the aqueous phase. The nitrated aromatic then diffuses back into the organic phase. Hanson et al. (1971) state that the nitration reaction occurs in the aqueous phase. The proposed mechanism is

$$HNO_3 + 2H_2SO_4 \leftrightarrow NO_2^+ + H_3O^+ + 2HSO_4^- \tag{3.15}$$

$$\mathbf{ArH} + NO_2^+ \leftrightarrow ArHNO_2^+ \tag{3.16}$$

$$ArHNO_2^+ + H_2O \leftrightarrow \mathbf{ArNO_2} + H_3O^+ \tag{3.17}$$

The species in bold type distribute preferentially into the organic phase.

The ArH in the nonpolar phase reacts with the NO_2^+ in the polar phase to form $ArHNO_2^+$, which resides in the polar phase. This reacts with water to form $ArNO_2$, which is soluble in the nonpolar phase.

Aromatic nitration kinetics relies upon the reaction kinetics and mass transfer kinetics. The former depends on the temperature and the compositions of both phases. The latter depends on the driving force, interfacial area, and the mass transfer coefficient. The area and coefficient depend on the power input per unit volume, the dispersed phase hold-up, the dispersed phase, the viscosity and density of both phases, the interfacial tension, and the molecular diffusivities of the solutes in each phase.

3.3.8 UOP HF Alkylation Technology

This process catalytically combines light olefins, which are typically mixtures of propylene and butylenes, with isobutane to produce branched-chain paraffinic fuel. The HF Alkylation[TM] process was developed by UOP during the late 1930s and early 1940s. The importance of the HF Alkylation process increased in the 1990s, according to Sheckler and Shah (1996, p. 1.32). In this process, hydrofluoric (HF) acid is used to catalyze the isoparaffin–olefin reaction, and the primary reactions are

$(CH_3)_2C=CH_3$ + $CH_2CH(CH_3)_2$ → $(CH_3)_3CCH_2CH(CH_3)_2$ (3.18)
   isobutylene   isobutane   2,2,4-trimethylpentane

$CH_2=CH-CH_2CH_3$ + $(CH_3)_3CH$ → $(CH_3)_2CHCH(CH_3)CH_2CH_2CH_3$
  1-butene   isobutane   2,3-dimethylhexane

                                                       (3.19)

$CH_3CH=CHCH_3$ + $(CH_3)_3CH$ → $(CH_3)_3CCH_2CH(CH_3)_2$
  2-butene   isobutane   2,2,4-trimethylpentane

or $CH_3CH(CH_3)CH(CH_3)CH(CH_3)_2$ (3.20)
      2,3,4-trimethylpentane

$CH_3CH=CH_2$ + $(CH_3)_3CH$ → $(CH_3)_2CHCH(CH_3)CH_2CH_3$ (3.21)
  propylene   isobutane   2,3-dimethylpentane

These reactions comprise an initiation step, a propagation step, an isomerization step, and possible polymerization and cracking steps. The mechanisms of these steps are given by Sheckler and Shah (1996, pp. 1.34–1.35). The initiation step involves the reaction of the olefin (nonpolar phase) with HF (polar phase) to form a fluorinated hydrocarbon, and a tertiary butyl cation that will subsequently carry the reaction. The propagation reaction involves the tertiary butyl cation reacting with an olefin to form a larger carbenium ion, which abstracts a hydride from an isobutane molecule. This reaction produces the product isoparaffin and a new tertiary butyl cation to continue the chain reaction. There are a number of secondary reactions, including hydrogen transfer, polymerization, isomerization, and destructive alkylation that result in the formation of secondary products that are both lighter and heavier than the primary products. A simplified flow diagram of the HF Alkylation unit for C_3–C_4 feed is shown in Figure 3.6. Treated and dried olefinic feed and recycle and makeup isobutane are fed to the reactor through a series of nozzles along its length to maintain an even temperature throughout. The reactor effluent is sent to the settler, where the acid is recycled to the reactor.

The nonpolar phase enters the isostripper along with saturate field butane feed. Product alkylates exit the base of the isostripper, and any normal butane that may have entered the process is removed via a sidedraw and recycled to the reactor. The isostripper overhead comprises isobutane, propane, and HF acid. The unit is primarily constructed of carbon steel, although Monel is used in several locations. In the reactor design, the critical design parameters are: (1) removing the heat of reaction, (2) mixing of the two-liquid-phase dispersion, (3) acid composition, and (4) introduction of the olefin feed.

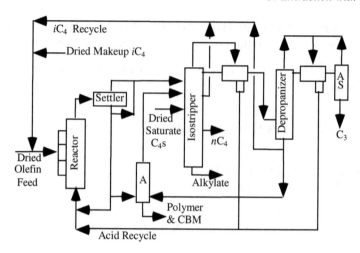

Figure 3.6: UOP C_3–C_4 HF alkylation process. A = HF acid regenerator; CBM = constant boiling mixture; AS = HF stripper. Products are treated with KOH.

3.3.9 UOP HF Detergent Alkylate Process

The detergent industry originated in the 1940s with the advent of sodium alkylbenzene sulfonates, which had detergent properties that were far superior to those of natural soaps. These soaps were derived by the saponification of naturally occurring triglycerides from either animal or vegetable origin to produce the sodium salts of the free fatty acids. Originally, the alkylbenzene sulfonates were essentially sodium dodecylbenzene sulfonates (DDBS), also known as branched alkylbenzene sulfonates (BAS). Their low cost and high effectiveness allowed them to quickly displace natural soaps in household applications. However, it was determined in the 1960s that the BAS exhibited slow biodegradation rates, so linear alkylbenzene sulfonates (LAS or LABS) started to displace the BAS because the former exhibited degradation rates that were similar to that of natural soaps. By the middle 1990s, LAS accounted for essentially the entire worldwide production of alkylbenzene sulfonates, and their annual production capacity is expected to grow to 2.7 million metric tons per year (Pujado, 1997, p. 1.53). There are several routes to produce linear alkylbenzenes (LAB), which include the alkylation of linear olefins with benzene via HF catalyst (see above). Alpha-olefins from ethylene oligomerization (Albemarle, Chevron, and Shell) or linear internal olefins from olefin disproportionation (SHOP^{TM} process) are used.

UOP has developed two methods to effect the alkylation: the HF Detergent Alkylate™ route that utilizes a liquid HF catalyst, and the new Detal™ process, developed jointly with PETRESA, which uses a fixed bed of acidic noncorrosive catalyst to replace the liquid HF. The corrosive nature of the liquid HF catalyst and the associated metallurgical requirements provide the solid catalyst an economic advantage. The HF Detergent Alkylate process is shown in Figure 3.7, where the linear paraffin–olefin feed is combined with makeup and recycle benzene and is cooled prior to being mixed with the HF acid feed. The reactor is mixed and a settler follows. Some of the acid from the settler is sent to the acid regenerator for the removal of heavy alkylate byproducts, and some is directly recycled to the alkylator. The light hydrocarbon phase from the settler is sent to the fractionation section, where the remaining HF acid, unreacted benzene and *n*-paraffins, heavy alkylate, and the LAB product are separated via fractionation columns. Linear paraffins (LP) and linear detergent alkylate product (LDA) are separated from the heavy alkylate. The benzene and HF acid are recycled to the reactor, and the *n*-paraffins are passed through alumina to remove combined fluorides before the stream is recycled to the dehydrogenation unit. The HF acid handling and acid neutralization equipment is also shown as it contributes significantly to the cost of the alkylation plant.

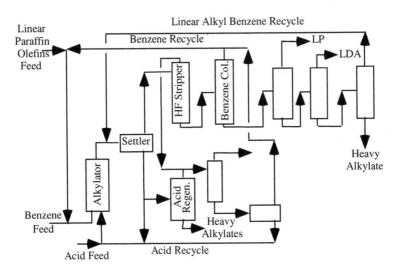

Figure 3.7: UOP HF Detergent Alkylate™ process. LP = linear paraffin; LDA = linear detergent alkylate product.

3.3.10 Oxidation

The UOP Merox^{TM} process is a catalytic process for the removal of sulfur, in the form of mercaptans, from petroleum fractions to improve gasoline stocks, to reduce odor from gasoline, and to meet sulfur specifications of liquid petroleum gas (LPG). The first Merox process unit was brought on-line on October 20, 1958, and in October 1993, the 1500th Merox process unit was commissioned (Holbrook, 1997, p. 11.37). Reactive extraction is used for Merox extraction and Merox sweetening. The former is used when low-molecular-weight mercaptans are transferred from the LPG or gasoline into an aqueous caustic soda solution to reduce the sulfur content of the product. The caustic soda solution and the H_2S-free feed enter the top and bottom of the countercurrent extraction column, respectively, as shown in Figure 3.8(a). The mercaptan-rich caustic solution (Rich Merox Caustic) exits the base of the extractor and flows through a steam heater, which is used to maintain the oxidizer temperature. The Rich Merox Caustic, Merox catalyst, and air flow upward through the oxidizer, where the caustic is regenerated by converting the mercaptans to disulfides. The oxidizer effluent then enters a decanter, where the disulfide oil is separated from the regenerated caustic solution. The process chemistry for the Merox extraction is focused on the mercaptan oxidation, which includes the following two steps:

$$\mathbf{RSH} + NaOH \rightarrow NaSR + H_2O \tag{3.22}$$

$$4NaSR + O_2 + 2H_2O \xrightarrow{\text{Merox catalyst}} 4NaOH + 2\mathbf{RSSR} \tag{3.23}$$

The species in bold type distribute preferentially into the organic phase.

Merox sweetening can be accomplished by a number of technologies including liquid–liquid sweetening, see Figure 3.8(b). Although this process

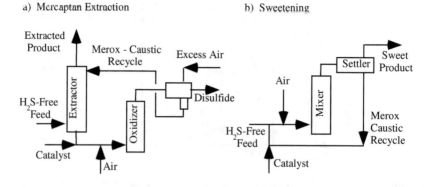

Figure 3.8: UOP Merox processing.

is reportedly not generally used for new units (Holbrook, 1997, p. 11.33), it is applicable to virgin light, thermally cracked gasolines, and to components with endpoints up to 120°C. In the sweetening process, the mercaptans are difficult to extract, and the hydrocarbon, Merox catalyst, air, and caustic solution are contacted in a cocurrent upflow reactor. The effluent from the reactor is decanted, and the aqueous caustic solution, which contains the Merox catalyst, is recycled to the reactor. It is believed that the thiol ($-$SH) group first transfers into the aqueous caustic solution, where it combines with the Merox catalyst. This complex oxidizes in the presence of oxygen to form the disulfide molecule in the water. These reactions are:

$$4RSH + O_2 \xrightarrow{\text{Merox catalyst and alkalinity}} 2RSSR + 2H_2O \qquad (3.24)$$

$$2R'SH + 2RSH + O_2 \xrightarrow{\text{Merox catalyst and alkalinity}} 2R'SSR + 2H_2O \qquad (3.25)$$

The latter equation represents the case where two different mercaptans enter into the reaction.

3.3.11 Oximation

Worldwide manufacturing capacity for ε-caprolactam was 3.7 million tons in 1995 (Weissermel and Arpe, 1997, p. 251), and about 85% of the world's caprolactam is based on two classical intermediates: cyclohexanone and cyclohexanone oxime (Simons and Haasen, 1991, p. 557). The classical manufacture of ε-caprolactam involves three "organic" steps and one "inorganic" step: (1) manufacture of cyclohexanone, (2) oximation of cyclohexanone with hydroxylamine, (3) Beckmann rearrangement of the cyclohexanone oxime to ε-caprolactam, and (4) production of hydroxylamine. The classical manufacture is widely used, but it has an inherent problem: about 4.4 pounds of ammonium sulfate is produced per pound of caprolactam; the contribution to this total for each step is summarized in Table 3.4.

DSM improved the oximation step in the classical process by using their hydroxylamine–phosphate–oxime (HPO) process; see Figure 3.9. In the inorganic recycle loop, ammonia and air are fed to a combustor that produces NO_3^-. The NO_3^- joins the aqueous-phase recycle before entering the "hyam" (hydroxylamine, or NH_3OH^+) reactor that is considered to be the fourth step listed above. In this reactor, hydroxylamine is produced by the reduction of nitrate ions with hydrogen in the presence of a noble metal catalyst and a phosphate buffer

Table 3.4:   Sources of Ammonium Sulfate in the Production of ε-Caprolactam by the Classical Manufacture Process.

Step	Pounds of Ammonium Sulfate Produced per Pound of Caprolactam
2: Oximation of cyclohexanone	1.1
3: Beckmann rearrangement	1.7
4: Production of hydroxylamine	1.6
Total	4.4

$$NO_3^- + 2H^+ + 3H_2 \xrightarrow{\text{Pd on carbon or Al}_2O_3 \text{ catalyst/PO}_4^{3-}} NH_3OH^+ + 2H_2O$$

$$(3.26)$$

The effluent from the hyam reactor is countercurrently contacted in a series of mixer–settlers with a toluene-rich fluid that contains cyclohexanone. In these extractors, the cyclohexanone, in the toluene phase, reacts with the hydroxylamine, in the aqueous phase, to form cyclohexanone oxime by the following reaction:

$$C_6H_{10}(=O) + NH_3OH^+ + 2H_2PO_4^- \rightarrow$$
$$C_6H_{10}(=NOH) + H_2PO_4^- + H_3PO_4 + H_2O \quad (3.27)$$

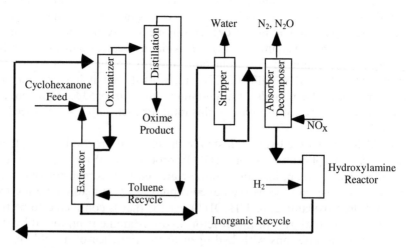

Figure 3.9:   DSM hydroxylamine–phosphate–oxime process.

Acid liberated during the oximation does not need to be neutralized, since the production of hydroxylamine and oximation occur in a recirculating buffer solution of phosphoric acid and ammonium dihydrogen phosphate. Therefore, the formation of ammonium sulfate is eliminated by the DSM^{TM} process in the second and fourth steps (see above). The cyclohexanone oxime distributes into the toluene phase, and the oxime is separated from the toluene via distillation. The toluene is recycled to the base of a counter-current extractor to recover any oxime and cyclohexanone that remained in the aqueous phase after the oximation reactor. A pulsed packed column may be used for this separation. The oxime that leaves the toluene–oxime distillation column is subjected to Beckmann rearrangement, in which it is rearranged to caprolactam in an acid medium. The resulting reaction product is a viscous mixture of caprolactam and oleum that is a single liquid phase. The separation of the caprolactam from this homogeneous stream can be accomplished by neutralizing it to a pH of about 4.5 with ammonia and allowing the two resulting immiscible phases to separate. Further processing steps are described by Simons and Haasen (1991, pp. 560–565) to produce caprolactam.

3.3.12 Shell Higher Olefin Process (SHOP)

Linear alkylbenzene sulfonate (LAS)-based detergents have essentially replaced branched alkylbenzene sulfonates (BAS), since the biodegradation rates of the former are closer to that of natural soaps than the latter. LAS accounts for essentially all of the linear alkylbenzene (LAB) worldwide production (Pujado, 1997, p. 1.53), and the demand for LAB has increased from 1 to 1.7 million metric tons per year from 1980 to 1990. Fatty alcohols from natural sources provide chain lengths from C_4–C_{18}, and the Ziegler method of oligomerizing ethylene produces a wide distribution of molecular weight products. However, the SHOP process provides an excellent method for producing the desired chain length material. This process was developed at Shell Research Company in the late 1960s and early 1970s, as disclosed by Baur et al. (1972a–f), Van Zwet et al. (1972), and Glockner et al. (1972). The first commercial plant was built in Geismar, LA, in 1977 with a capacity of 200,000 metric tons per year and that was expanded to 590,000 metric tons by 1992 (Vogt, 1996, p. 247). A second plant utilizing this technology was commissioned in Stanlow, UK, with the capacity by 1992 of 278,000 metric tons. With a total capacity of nearly one million metric tons, the SHOP process is one of the largest applications of homogeneous catalysis by a transition metal, according to Keim (1990).

The SHOP process combines ethylene oligomerization, isomerization, and metathesis to produce α-olefins for detergents, and internal olefins.

The oligomerization reaction is performed in a polar solvent in which a nickel catalyst is dissolved. The α-olefins are nearly insoluble in the polar solvent, and therefore produce a second immiscible liquid phase. This is the key feature of the process, according to Vogt (1996, p. 251), and was one of the first examples of a biphasic liquid–liquid system to be used in catalysis. The preferred polar solvents are alkanediols, especially 1,4-butanediol. The catalyst is formed in situ by combining a nickel salt and a chelating ligand (like o-diphenylphosphinobenzoic acid). The oligomerization is performed at 80–140°C and at 7–14 MPa, and it produces up to 99% linearity and 96–98% terminal olefins over the range C_4–C_{30+}. The overall process flowsheet is presented in Figure 3.10, and involves two reactors in series that feed a decanter. The more dense polar catalytic solution is recycled to the reactor, and the less dense nonpolar α-olefin phase is washed. The α-olefins are sent to a series of distillation columns, where the desired product fractions are recovered. The C_4–C_{10} and C_{20+} α-olefin fractions are combined and iso-merized to internal linear olefins, which are fed to metathesis reactors that produce a mixture of olefins with odd and even carbon chain lengths. This mixture is fractionated via distillation to produce the desired C_{11}–C_{14} linear internal olefin cuts, and the other cuts may be recycled to the metathesis reactor. The mechanism of the reaction is described by Vogt (1996, pp. 253–258), Keim (1990), and Keim et al. (1986).

3.3.13 Hydroformylation

The reaction between olefinic double bonds and a mixture of hydrogen and carbon monoxide leads to linear and branched aldehydes, which may in turn be converted to a wide bandwidth of compounds, including alcohols, amines, and acids. The primary markets for alcohols produced by hydro-

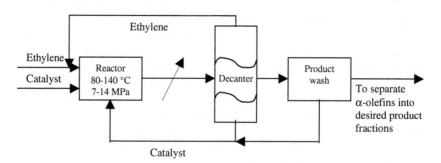

Figure 3.10: Flow diagram of the Shell Higher Olefin Process (SHOP). (After Vogt, 1996, p. 252.)

formylation are in polyvinyl chloride (plasticizers) and detergent industries (Allen et al., 1990, pp. 63–67). The first-generation hydroformylation process was based on cobalt at rather extreme conditions of between 200 and 350 bar and between 150 and 180°C. Since the reaction conditions were quite similar between competitors (BASF, ICI, Ruhrchemie, and Kuhlmann), the processes differed by the method of product–catalyst separation, according to Frohning and Kohlpaintner (1996, p. 31). Shell researchers discovered the utility of phosphines, which allowed the reaction to occur at lower carbon monoxide pressure. Therefore, the Shell process that utilized cobalt–phosphine catalyst was the evolutionary result of the first-generation hydroformylation process. The second-generation hydroformylation process utilized a ligand (such as triphenylphosphine, TPP) modified rhodium catalyst, which had both material and energy utilization advantages over the cobalt-based technology. Subsequent to 1976, the Union Carbide Corporation (UCC) began to license the low-pressure oxo (LPO^{TM}) process for the hydroformylation of propylene. A flowsheet of this process with liquid-phase recycle is given in Figure 3.11. Note that the product–catalyst separation is effected by vaporization of the product. An alternative hydroformylation process was developed in the early 1980s with innovations in reaction engineering that utilized water-soluble phosphine ligands (such as trisulfonated triphenylphosphine, TPPTS) in a biphasic reaction system. Industrial hydroformylation is currently effected by two basic processes that involve the catalyst and substrate in the same liquid phase (UCC, BASF, and Shell) and in two-liquid phases, with a water-soluble catalyst (Ruhrchemie/Rhone-Poulenc).

The Ruhrchemie/Rhone-Poulenc biphasic process for the hydroformylation of propylene was commercialized in 1984, and the plant has subsequently been expanded to a capacity of more than 300,000 tons per year

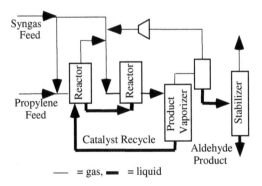

Figure 3.11: Low-pressure oxo hydroformylation process.

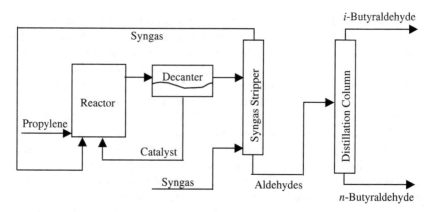

Figure 3.12: Ruhrchemie/Rhone-Poulenc process for the hydroformylation of propylene using water-soluble rhodium complexes. A decanter is used to separate the products from the homogeneous catalyst.

(Frohning and Kohlpaintner, 1996, p. 80). A unit to convert *n*-butene to *n*-pentanal was commercialized in 1995. The flowsheet for this process (Figure 3.12), includes a liquid–liquid–vapor phase reactor, a decanter and degasser for the crude aldehyde product, a heat exchanger for the aqueous catalyst, which produces steam, a stripping column where syngas is used to strip unreacted olefin, and a conventional distillation column to effect the iso–normal isomer separation. There are several differences between this system and LPO. First, the reaction is relatively insensitive to sulfur and other oxo poisons that may enter with the feed. This reduces the cost of feed pretreatment. Second, organic-soluble byproducts are purged through the vent stream and with the product stream. A comparison of the liquid–liquid–gas and the LPO processes was summarized in Frohning and Kohlpaintner (1996, pp. 79, 82) and Billig and Bryant (1996).

3.3.14 Modifiers, Phase Transfer Catalysts, and Surfactants

A phase transfer agent is used in catalytic quantities to effect interfacial species transfer and to facilitate reactions between reagents in two immiscible phases. The industrial use of phase transfer catalysis (PTC) has increased from less than 10 applications in 1975 to more than 600 in the early 1990s, which manufacture products valued at between 10 and 20 billion dollars, according to Starks et al. (1994, p. xiii). Applications of PTC have occurred in many industries including pharmaceutical, chemical, agricultural, and perfumery and fragrance.

PTC reactions are applicable to a vast number of systems, including nucleophilic substitution reactions and reactions in the presence of bases, which involve the deprotonation of moderately and weakly acidic organic compounds. PTC is commonly utilized in a multistep industrial synthesis of a fine chemical, which could involve displacement, polymerization, oxidations, reductions, chiral, and transition metal co-catalyzed reactions. The books by Starks et al. (1994), Dehmlow and Dehmlow (1993), and Weber and Gokel (1977) are extremely valuable resources for a broad range of reactions involving PTC.

A classic example of PTC was described by Starks et al. (1994, p. 1) and is illustrated in Figure 3.13 for the reaction of 1-chlorooctane (without a solvent) and aqueous sodium cyanide. Heating the two-phase mixture in the absence of catalyst at reflux conditions with vigorous agitation for 1–2 days results in no apparent reaction, with the possible exception that some of the sodium cyanide was hydrolyzed to ammonia and sodium formate. However, upon the addition of 1 wt% of an appropriate quaternary ammonium salt (PTC), say $(C_6H_{13})_4N^+Cl^-$, to the mixture, the displacement reaction occurred rapidly, producing 1-cyanooctane in near 100% selectivity and 100% conversion in 2–3 hours.

Classification of PTC reactions is based on the solubility of the PTC in the liquid phase. Depending on the actual phases present, the reactions are subclassified as liquid–liquid (LLPTC), gas–liquid (GLPTC), and solid–liquid (SLPTC). Naik and Doraiswamy (1998, p. 613) note: " . . . the main disadvantage of PTC, especially in commercial applications, is the need to separate the catalyst from the product organic phase." A similar problem has previously been discussed in the hydroformylation section, and analogously, the placement of the catalyst in a separate phase is advantaged from a catalyst–product separation viewpoint. Therefore, PTC may be either immobilized on a solid support or placed in a third immiscible PTC-rich phase.

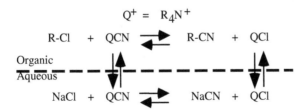

Figure 3.13: Classic representation of cyanide displacement by phase transfer catalysis.

Table 3.5: Selected Examples of Some Industrially Important PTC Reactions.

Application	Reaction	Reference
Chiral synthesis using cinchinidium-derived optically active phase transfer catalyst:		
Synthesis of indacrinone, a diuretic drug candidate	C-alkylation of indanone derivatives and oxyindoles using cinchoma alkaloids	Dolling et al. (1987) Bhattacharya et al. (1986)
Synthesis of chiral α-amino acids	Alkylation of imines, glycine derivatives, and Schiff base derivatives	O'Donnell (1993) O'Donnell et al. (1989)
Polymerization reactions:		
Condensation reactions	Synthesis of polycarbonates, polyesters, polysulfonates, and polyethers	Tagle et al. (1994) Leung et al. (1994)
Free radical polymerizations	Polymerization of acrylonitrile using potassium peroxomonosulfate as initiator	Balakrisnan and Arivalagan (1994)
Anionic polymerizations	Diene polymerization in the presence of crown ethers	Reetz and Ostarek (1988) Cheng (1984)
Chemical modifications of polymers	Modifications of chloromethyl-substituted polystyrene and poly(vinyl halides)	Fréchet (1984)
Agrochemicals:		
Synthesis of an antidote for herbicides	N-alkylation of hexamethylenetetraamine with chloromethyl ketones	Smith (1990)
Synthesis of herbicides and insecticides	Selective O-alkylation and O-phosphorylation of ambient pyridinates	Cutie and Halpern (1992)
Synthesis of insecticidal pyrethroid and insect pheromones	Wittig reaction of aliphatic aldehydes and alkenyl alcohols with 50% NaOH or solid K_2CO_3	Deng et al. (1989)
Synthesis of naturally occurring pellitorine, possessing insecticidal activity	PTC vinylation of (E)-1-iodo-1-heptene with vinyl acetate	Jeffrey (1988)

Synthesis of a herbicide	Fujita et al. (1982)
Synthesis of an intermediate for the preparation of insecticidal pyrethroids	Galli et al. (1984)

N-alkylation of substituted phenyl N-hydroxyurea with dimethyl sulfate

PTC Wittig reaction of *trans*-caronaldehyde ethyl ester with 50% NaOH and an in situ-generated PT catalyst

Perfumery and fragrance industry:

Enhancement and augmentation of aroma of perfumes	Sprecker and Hanna (1982)
Intermediate step in the synthesis of a fragrance from furfural	Norwicki and Gora (1991)
Synthesis of phenylacetic acid, an intermediate in the perfumery industry	Cassar et al. (1976)

Alkylation of acetophenone moiety with allyl chloride

C-alkylation of propanal and butanal by 2-chloromethylfuran

Carbonylation of benzyl chloride in the presence of a palladium-based catalyst

Compounds with biological activity:

One-pot synthesis of carboxamides and peptides	Watanabe and Mukiyama (1981a)
Synthesis of intermediates in nucleic acid chemistry	Grouiller et al. (1987)
One-pot synthesis of benzofuran derivatives, with wide ranging biological activities	Ayyangar et al. (1987)
Synthesis of aminopyrroles, intermediates in synthesis of biologically active compounds like pyrrolyltriazenes	Almerico et al. (1989)

Reaction of a free acid or a carboxylic ester with an amine with KOH/K₂CO₃ and a phenylphosphonate coupling agent

Regioselective synthesis of *p*-toluenesulfonyl derivatives of carbohydrates and nucleosides

Reaction of *o*-chloronitrobenzenes with sodium azide

N-alkylation of N-unsubstituted 3-aminopyrrole with TDA-1 as PT catalyst

(continued)

Table 3.5 (continued)

Application	Reaction	Reference
Pharmaceuticals:		
Synthesis of various drugs like dicyclonine, phenoperidine, oxaladine, ritaline, etc.	Alkylation of phenylacetonitrile using NaOH, instead of expensive sodium ethoxide	Lindbloom and Elander (1980)
Synthesis of (R)-fluorenyloxyacetic acid, useful in the treatment of brain edema	Use of a nonionic surfactant, Triton X, with a cinchonidinium-based PT catalyst to accelerate the alkylation step	Dolling (1986)
Synthesis of commercial antibiotic, chloramphenicol	Aldol condensation in the presence of NaOH and a PT catalyst	Koch and Magni (1985)
Synthesis of penicillin-based compounds (Astra AB, Sweden)	Selective esterification of benzylpenicillin using α-chloroethyl carbonate	Lindblom and Elander (1980)
Synthesis of chloropromaine and imipramine, an antidepressant	N-alkylation of carbazones, phenothiazines, acridanone, and indoles using alkyl halides and aqueous NaOH/solid K_2CO_3	Schmolka and Zimmer (1984)
Synthesis of lysergic acid-based pharmaceuticals and other molecules with the indole skeleton	Facile and selective monoalkylation of the indole nitrogen using PTC, instead of using K-azide in liquid ammonia at −40°C	Lindblom and Elander (1980)
Synthesis of calcitriol derivatives	O-alkylation using *tert*-butylbromoacetate	Neef and Steinmeyer (1991)
Synthesis of drugs and pharmacologically active agents	N-alkylation of phenothiazines, carboamides, and β-lactams	Masse (1977)
Other specialty chemicals:		
Synthesis of chloroprene	Dehydrohalogenation of 3,4-dichloro-1-butene	Maurin (1983)
Synthesis of allyltribromophenol, a flame retardant polymer	Etherification of hindered tribromophenol with allyl bromide	Wang and Yang (1990)
Synthesis of prepolymers based on natural resources like lignin	Reaction of hydroxylalkyl modified lignin with epichlorohydrin and solid KOH	Glasser et al. (1990)

Other specialty chemicals: (*continued*)

Synthesis of dialkyl sulfides (additives for lubricants, stabilizers for photographic emulsions)	Reaction of sodium sulfide with benzyl chloride	Pradhan and Sharma (1992); Hagenson et al. (1994)
Synthesis of spiro derivatives of tetrahydrothiophene, a characteristic fragment of many alkaloids	Spiro-linking of tetrahydrothiophene ring to a substituted quinolizidine skeleton	Wrobel and Hejchman (1987)
Synthesis of β-lactams	Reaction of amino acids and methanesulfonyl chloride	Watanabe and Mukiyama (1981b)
Synthesis of dichlorovinyl carbazole, used in preparation of photoconductive polymers	Dichlorovinylation of carbazole in solid–liquid system	Pielichowski and Czub (1995)
Synthesis of macrolide-like lactones	Synthesis of lactones from the conjugate base of ω-bromocarboxylic acids	Kimura and Regen (1983b)
Synthesis of dyes derived from desyl esters	Reaction of desyl alcohol with NaOH, chloroform and a PT catalyst, followed by a PEG catalyzed chloride displacement	Shenoy and Rangnekar (1989)

Source: Naik and Doraiswamy (1998, pp. 638–639).

PTC agents include: ammonium, phosphonium, and other onium salts; macrocyclic polyethers (crown ethers), aza-macrobicyclic ethers (cryptands); and open-chain polyethers including polyethylene glycols (PEG), their dimethyl ethers, and glymes. Quaternary onium salts are the most frequently used phase transfer agents. Crown ethers and cryptands are also used, especially in SLPTC, due to their ability to complex and solubilize metal cations and the associated anion. However, the cost and toxicity of crown ethers and cryptands hinder their industrial utility. Although PEGs are less active than quaternary ammonium salts and crown ethers, they are less costly and are environmentally safe. They are stable, biodegradable, easy to recover, and nontoxic. However, their substantial aqueous solubility limits their utility in liquid–liquid systems, although in some cases PEGs form a third catalyst-rich phase that functions as an active catalyst.

Many industrially important reactions involve the use of PTC in the presence of a base such as aqueous NaOH or K_2CO_3. These include C-, N-, O-, S-alkylations, isomerizations, additions, β- and α-eliminations, and hydrolysis reaction. An advantage of performing reactions in the presence of base is the prevention of hydrolysis of the organic reactant, since the organic substrate is not subjected to the alkaline conditions present in the aqueous phase and OH^- has only limited solubility in the organic phase.

Table 3.5 summarizes typical representative reactions that are commercially practiced (Naik and Doraiswamy, 1998, pp. 638–639).

3.3.15 Dense Gas and Ionic Fluid Applications

Both liquid and supercritical carbon dioxide, other dense gases, and ionic fluids provide alternative solvents whose properties present potentially favorable conditions for reactive extraction processes. Carbon dioxide is relatively immiscible with water and thus provides an alternative to hydrocarbon solvents. Ionic fluids can be immiscible with both water and hydrocarbon phases, and they are also alternative solvents. Research has been performed using carbon dioxide as a reaction medium in place of traditional organic solvents (see reviews by McHugh and Krukonis, 1994, and by Brunner, 1994). Dehghani et al. (1996) discussed the use of carbon dioxide to replace hydrocarbon solvents for metal extraction, and Froschl and Marr (1993) examined the use of amine entrainers with supercritical fluids. Many alternative reactions have been performed including polymerization and hydrolysis. Likewise, experimentation using supercritical water for oxidation has been performed. Ionic fluids present a new possibility (Huddleston et al., 1998; Visser et al., 2000a,b).

3.4 Conclusions

This chapter has described industrial reactive extraction operations. Although it has covered the reversible reaction processes using hydrometallurgical examples and irreversible reaction processes using organic examples, both irreversible and reversible reactions can be combined. Likewise, fractional extraction can be used to intensify both the reaction and separation in the same cascade and by providing zones for reaction and zones for separation in adjoining cascades. The combination of these concepts has rarely been performed. The PUREX process combines all of these concepts; however, each element of it is rarely considered outside the nuclear industry. It is the hope of the authors that by using both inorganic and organic processes together, technology transfer will be promoted and new processes will emerge.

References

Albright, L.F., "Nitration," in *Kirk–Othmer Encyclopedia of Chemical Technology*, 4th Edition, Wiley & Sons, New York, **17**, 1996.

Allen, P.W., R.L. Pruett, and E.J. Wickson, "Oxo Process Alcohols," in *Encyclopedia of Chemical Processing and Design*, J.J. McKetta and W.A. Cunningham, Eds., Marcel Dekker, Inc., New York, **33**, 1990.

Almerico, A.M., G. Cirrinclone, E. Aiellio, and G. Daltolo, *J. Heterocycl. Chem.*, **26**, 1631, 1989.

Al-Saadi, A.N. and G.V. Jeffreys, *AIChE J.*, **27**, 761, 1981a.

Al-Saadi, A.N. and G.V. Jeffreys, *AIChE J.*, **27**, 768, 1981b.

Al-Saadi, A.N. and G.V. Jeffreys, *AIChE J.*, **27**, 754, 1981c.

Alwan, S. et al., *Chem. Eng. Commun.*, **22**, 317, 1983.

Aslam, M., G.P. Torrence, and E.G. Zey, "Esterification," in *Kirk–Othmer Encyclopedia of Chemical Technology*, 4th Edition, Wiley & Sons, New York, **4**, 1994.

Ayyangar, N.R., S. Madan Kumar and K.V. Srinivasan, *Synthesis*, No. 7, 616–618, 1987.

Balakrishnan, T. and K. Arivalagan, *J. Polym. Sci., Part A: Polym. Chem.*, **32**, 1909, 1994.

Baur, R., H. Chung, K.W. Barnett, P.W. Glockner, and W. Keim, U.S. Patent 3,686,159 to Shell Dev. Co., 1972a.

Baur, R., H. Chung, D. Camel, W. Keim, and H. van Zwet, U.S. Patent 3,637,636 to Shell Dev. Co., 1972b.

Baur, R., H. Chung, W. Keim, and H. van Zwet, U.S. Patent 3,661,803 to Shell Dev. Co., 1972c.

Baur, R., P.W. Glockner, W. Keim, H. van Zwet, and H. Chung, U.S. Patent 3,644,563 to Shell Dev. Co., 1972d.

Baur, R.S., P.W. Glockner, W. Keim, and R.F. Mason, U.S. Patent 3,647,915 to Shell Dev. Co., 1972e.

Baur, R.S., H. Chung, P.W. Glockner, W. Keim, and H. van Zwet, U.S. Patent 3,635,937 to Shell Dev. Co., 1972f.

Bhattacharya, V.A., U. Dolling, E.J. Grabowski, S. Krady, K.M. Ryan, and L.M. Weinstock, *Angew. Chem., Int. Ed. Engl.,* **25**, 476, 1986.

Bhave, R.R. and M.M. Sharma, *Trans. Inst. Chem. Eng.,* **59**, 161–169, 1981.

Billig, E. and D.R. Bryant, "Oxo Process," in *Kirk–Othmer Encyclopedia of Chemical Technology,* 4th Edition, Wiley & Sons, New York, **17**, 1996.

Blumberg, R., *Liquid–Liquid Extraction,* Academic Press, New York, 1988.

Booth, G., "Nitro Compounds, Aromatic," in *Ullmann's Encyclopedia of Industrial Chemistry,* 5th Completely Revised Edition, B. Elvers, S. Hawkins, and G. Schulz, Eds., VCH Verlagsgesellschaft mbH, Federal Republic of Germany, **A17**, 1991.

Brockmann, R., G. Demmering, U. Kreutzer, M. Lindemann, J. Plachenka, and U. Steinberner, "Fatty Acids," in *Ullmann's Encyclopedia of Industrial Chemistry,* W. Gerhartz, Ed., VCH Verlagsgesellschaft mbH, Germany, **A10**, 1987.

Brunelle, D.J., "Polycarbonates," in *Kirk–Othmer Encyclopedia of Chemical Technology,* 4th Edition, Wiley & Sons, New York, **9**, 1996.

Brunner, G., *Gas Extraction: An Introduction to Fundamentals of Supercritical Fluids and the Application to Separation Processes,* Steinkopff, Darmstadt; Springer, New York, 1994.

Buysch, H.J., U. Jansen, P. Ooms, E.G. Hoffmann, and B.U. Schenke, U.S. Patent 5,750,801, 1998.

Cassar, L., M. Foa and A. Gardano, *J. Organomet. Chem.,* **121**, C55, 1976.

Cheng, T.C., in *Crown Ethers and Phase-Transfer Catalysis in Polymer Science,* L.J. Mathias and C.E. Carraher, Jr., Eds., Plenum Press, New York, 1984, p. 155.

Converse, A.O. and H.E. Grethlein, U.S. Patent 4,556,430, 1985.

Cutie, Z.G. and M. Halpern, U.S. Patent 5,120,846, *Chem. Abstr.,* **117**, 90510c, 1992.

Davies, G.A. and G.V. Jeffreys, in *Recent Advances in Liquid–Liquid Extraction,* C. Hanson, Ed., Pergamon Press, New York, 1971, Chapter 14.

Dehghani, F., T. Wells, N.J. Cotton, and N.R. Foster, in *Value Adding Through Solvent Extraction,* D.C. Shallcross, R. Paimin, and L.M. Prvcic, Eds., University of Melbourne Press, Melbourne, **2**, 1996, p. 967.

Dehmlow, E.V. and S.S. Dehmlow, *Phase-Transfer Catalysts,* 3rd Edition, VCH Publishers, New York, 1993.

Deng, Y.N., H.D. Li, and H.S. Xu, *Chin. Sci. Bull.* (*Chem. Abstr.,* **114**, 23571y), **34**, 203, 1989.

Dolling, U.H., U.S. Patent 4,605,761, *Chem. Abstr.,* **106**, 4697n, 1986.

Dolling, U.H., D.L. Hughes, A. Bhattacharya, K.M. Ryan, S. Karady, L.M. Weinstock, V.J. Grenda, and E.J.J. Grabowski, in *Phase-Transfer Catalysis: New Chemistry, Catalysts, and Applications,* C.M. Starks, Ed., ACS Symposium Series (326), American Chemical Society, Washington, DC, 1987, p. 67.

Doraiswamy, L.K. and M.M. Sharma, in *Fluid–Fluid–Solid Reactions,* Wiley & Sons, New York, **2**, 1984.

Dorf, R., "Extraction," in *The Engineering Handbook*, CRC Press, Boca Raton, FL, 1996.

Dorner, M., U.S. Patent 5,744,624, April 28, 1998.

Fox, D.W., U.S. Patent 3,153,008, 1964.

Frantz, R.W. and V. Van Brunt, *Sep. Sci. Technol.*, **22**, 243, 1987.

Fréchet, J.M.J., in *Crown Ethers and Phase-Transfer Catalysis in Polymer Science*, L.J. Mathias and C.E. Carraher, Jr., Eds., Plenum Press, New York, 1984.

Frohning, C.D. and C.W. Kohlpaintner, in *Applied Homogeneous Catalysis with Organometallic Compounds*, B. Cornils and W.A. Herrmann, Eds., VCH, New York, 1996.

Froschl, F. and R. Marr, in *Solvent Extraction in the Process Industries*, Proceedings of ISEC'93, D.H. Logsdail and M.J. Slater, Eds., Elsevier Applied Science, New York, **2**, 1993.

Fujita, F., N. Itaya, H. Kishida, and I. Takemoto (Sumitomo Chemical Co. Ltd.), Eur. Patent Appl. 3835, U.S. Patent 4,328,166, *Chem. Abstr.*, **92**, 110707t, 1982.

Galli, R., L. Scaglioni, O. Palla, and F. Gozzo, *Tetrahedron,* **40**, 1523, 1984.

Glasser, W.G., W. DeOliveira, S.S. Kelly, and L.S. Nieh, U.S. Patent 4,918,167, *Chem. Abstr.*, **113**, 99698, 1990.

Glockner, P.W., W. Keim, and R.F. Mason, U.S. Patent 3,647,914 to Shell Dev. Co., 1972.

Groothuis, H. and F.J. Zuiderweg, *Chem. Eng. Sci.*, **12**, 288, 1960.

Grouiller, A., H. Essadiq, B. Najib, and P. Moliere, *Synthesis,* No. 12, 1121, 1987.

Gupta, S.K., U.S. Patent 5,468,887, 1995.

Hag, G.L.F. and R.K. Rantala, U.S. Patent 4,474,993, 1984.

Hagenson, L.C., S.D. Naik, and L.K. Doraiswamy, *Chem. Eng. Sci., ***49**, 4787, 1994.

Hanson, C., in *Recent Advances in Liquid–Liquid Extraction*, C. Hanson, Ed., Pergamon Press, New York, 1971.

Hanson, C., J.G. Marsland, and G. Wilson, *Chem. Eng. Sci.*, **26**, 1513, 1971.

Hanson, C., J.G. Marsland, and M.A. Naz, *Chem. Eng. Sci.*, **29**, 297, 1974.

Hiraoka, S., Y. Tada, H. Suzuki, H. Mori, T. Aragaki, and I. Yamada, *J. Chem. Eng. Japan*, **23**(4), 468, 1990.

Holbrook, D.L., in *Handbook of Petroleum Refining Processes*, 2nd Edition, R.A. Meyers, Ed., McGraw-Hill, New York, 1997, Chapter 11.3.

Huddleston, J.G., H.W. Willauer, R.P. Swatloski, A.E. Visser, and R.D. Rogers, *Chem. Commun.*, 1765–1766, 1998.

Jeffrey, T., *Synth. Commun.,* **18**, 77, 1988.

Jeffreys, G.V., V.G. Jenson, and F.R. Miles, *Trans. Inst. Chem. Eng.*, **39**, 389, 1961.

Jones, S.C. and D.G. Fallon, U.S. Patent 5,648,529, 1997.

Keim, W., *Angew. Chem., Int. Ed. Engl.*, **29**, 235, 1990.

Keim, W., A. Behr, B. Gruber, B. Hoffmann, F.H. Kowaldt, U. Kurschner, B. Limbacker, and F. Sistig, *Organometallics*, **5**, 2356, 1986.

Kimura, Y. and S.L. Regen, *J. Org. Chem.,* **48**, 1533, 1983.

Knaggs, E.A. and M.J. Nepras, "Sulfonation and Sulfation," in *Kirk–Othmer Encyclopedia of Chemical Technology*, 3rd Edition, Wiley & Sons, New York, **23**, 1982.

Koch, M. and A. Magni (Gruppo Lepetit), U.S. Patent 4,501,919, *Chem. Abstr.*, **102**, 204296k, 1985.

Koyama, Y., N. Kawase, H. Yamamoto, S. Kawata, and Y. Kasori, U.S. Patent 5,565,557, October 15, 1996.

Lawson, G. and G.V. Jeffreys, *Trans. Inst. Chem. Eng.*, **43**, 294, 1965.

Leung, L.M., W.H. Chan, S.K. Leung, and S.M. Fung, *J. Macromol. Sci.,-Pure Appl. Chem.*, **A31**, 495, 1994.

Lindblom, L. and M. Elander, *Pharmaceutical Tech.*, **4**, 59, 1980.

Lo, T.C., M.H.I. Baird, and C. Hanson, *Handbook of Solvent Extraction*, Wiley-Interscience, New York, 1983.

Maeda, S., A. Kondo, and S. Nishimura, U.S. Patent 3,993,699, 1976.

Marcus, Y. and A.S. Kertes, *Ion Exchange and Solvent Extraction of Metal Complexes*, Wiley-Interscience, New York, 1969.

Martin, J. and J.I. Krchma, U.S. Patent 1,770,414, 1930.

Masse, J., *Synthesis*, **5**, 341, 1977.

Maurin, L., U.S. Patent 4,418,232, *Chem. Abstr.*, **100**, 52179t, 1983.

McHugh, M.A. and V.J. Krukonis, *Supercritical Fluid Extraction: Principles and Practice*, 2nd Edition, Butterworth-Heinemann, Boston, 1994.

McKay, H.A.C., J.H. Miles, and J.L. Swanson, in *Science and Technology of Tributyl Phosphate*, W.W. Schultz and J.D. Navratil, Eds., CRC Press, Inc., Boca Raton, FL, **3**, 1991.

Minotti, M., Ph.D. Dissertation, University of Massachusetts, Amherst, 1998.

Morrison, R.T. and R.N. Boyd, *Organic Chemistry*, 3rd Edition, Allyn and Bacon, Inc., Boston, 1973.

Naik, S.D. and L.K. Doraiswamy, *AIChE J.*, **44**(3), 612, 1998.

Nanda, A.K. and M.M. Sharma, *Chem. Eng. Sci.*, **22**, 69, 1967.

Neef, G. and A. Steinmeyer, *Tetrahedron Lett.*, 5073, 1991.

Norwicki, J. and J. Gora, *Pol. J. Chem.*, **65**, 2267, 1991.

O'Donnell, M.J., in *Catalytic Asymmetric Synthesis*, I. Ojima, Ed., VCH Publishers, New York, 1993, p. 389.

O'Donnell, M.J., W.D. Bennett, and S. Wu, *J. Am. Chem. Soc.*, **111**, 2353, 1989.

O'Quinn, L.N. and V. Van Brunt, *Sep. Sci. Technol.*, **22**, 467, 1987.

Pielichowski, J. and P. Czub, *Synth. Commun.*, **25**, 3647, 1995.

Piret, E.L., W.H. Penny, and P.J. Trambouze, *AIChE J.*, **6**, 394, 1960.

Pradhan, N.C. and M.M. Sharma, *Ind. Eng. Chem. Res.*, **31**, 1610, 1992.

Pujado, P.R., in *Handbook of Petroleum Refining Processes*, R.A. Meyers, Ed., McGraw-Hill, New York, 1997, Chapter 1.5.

Reetz, M.T. and R. Ostarek, *J. Chem. Soc., Chem. Commun.*, **3**, 213, 1988.

Ridgway, K. and E.E. Thorpe, "Use of Solvent Extraction in Pharmaceutical Manufacturing Processes," in *Handbook of Solvent Extraction*, T.C. Lo, M.H.I. Baird, and C. Hanson, Eds., Krieger Publishing Company, Malabar, FL, 1991.

Ritcey, G.M. and A.W. Ashbrook, *Solvent Extraction, Principles and Applications to Process Metallurgy*, Elsevier, New York, **I**, 1984; **II**, 1979.

Sarkar, S., C.J. Mumford, and C.R. Phillips, *Ind. Eng. Chem. Process Des. Dev.*, **19**, 672, 1980.

Schmolka, S.J. and J. Zimmer, *Synthesis*, **1**, 29, 1984.

Schnell, H., L. Bottenbruch, and G. Grimm, U.S. Patent 3,028,365, 1962.

Schultz, W.W. and J.D. Navratil, *Science and Technology of Tributyl Phosphate*, CRC Press, Inc., Boca Raton, FL, **I**, 1984; **IIA**, 1987; **IIB**, 1987; **III**, 1990; **IV**, 1991.

Sekine, T. and Y. Hasegawa, *Solvent Extraction Chemistry, Fundamentals and Applications*, Marcel Dekker, New York, 1977.

Serini, V., "Polycarbonates," in *Ullmann's Encyclopedia of Industrial Chemistry*, 5th Completely Revised Edition, VCH Verlagsgesellschaft mbH, Federal Republic of Germany, **A21**, 1992.

Sharma, M.M., "Extraction with Reaction," in *Handbook of Solvent Extraction*, Reprint Edition, T.C. Lo, M.H.I. Baird and C. Hanson, Eds., Krieger Publishing Company, Malabar, FL, 1991.

Sharma, M.M. and A.K. Nanda, *Trans. Inst. Chem. Eng.*, **46**, T44, 1968.

Sheckler, J.C. and B.R. Shah, in *Handbook of Petroleum Refining Processes*, R.A. Meyers, Ed., McGraw-Hill, New York, 1996, Chapter 1.4.

Shenoy, G. and D. Rangnekar, *Dyes and Pigments*, **10**, 165, 1989.

Simons, A.F.J. and N.F. Haasen, in *Handbook of Solvent Extraction*, Reprint Edition, Krieger Publishing Company, Malabar, FL, 1991, Chapter 18.4.

Smith, A.R., J.E. Caswell, P.P. Lawson, and S.D. Cavers, *Can. J. Chem. Eng.*, **41**, 150, 1963.

Smith, L.R., U.S. Patent 4,962,212, *Chem. Abstr.*, **114**, 143116a, 1990.

Sprecker, M.A. and M.R. Hanna, U.K. Patent Appl. GBB2139222, U.S. Patent Appl. 487045, *Chem. Abstr.*, **102**, P220567e, 1982.

Starks, C.M., C.L. Liotta, and M. Halpern, *Phase-Transfer Catalysis: Fundamentals, Applications, and Industrial Perspective*, Chapman & Hall, New York, 1994.

Swanson, J.L., in *Science and Technology of Tributyl Phosphate*, W.W. Schultz and J.D. Navratil, Eds., CRC Press, Inc., Boca Raton, FL, **III**, 1991.

Tagle, L.H., F.R. Diaz, and R. Fuenzalida, *J. Macromol. Sci., Pure Appl. Chem.*, **A31**, 283, 1994.

Trambouze, P.J. and E.L. Piret, *AIChE J.*, **6**, 574, 1960.

Trambouze, P.J., M.L. Trambouze, and E.L. Piret, *AIChE J.*, **7**, 138, 1961.

Treybal, R.E., *Liquid Extraction*, 2nd Edition, McGraw Hill, New York, 1963.

Van Zwet, H., R. Baur, and W. Keim, U.S. Patent 3,644,564 to Shell Dev. Co., 1972.

Verrall, M.S., in *Science and Practice of Liquid–Liquid Extraction*, J.D. Thornton, Ed., Clarendon Press, Oxford, **2**, 1992.

Visser, A.E., R.P. Swatloski, S.T. Griffin, D.H. Hartman, and R.D. Rogers, *Sep. Sci. Technol.*, in press, 2000a.

Visser, A.E., R.P. Swatloski, and R.D. Rogers, *Green Chemistry*, **2**, 1–4, 2000b.

Vogt, D., in *Applied Homogeneous Catalysis with Organometallic Compounds*, B. Cornils and W.A. Herrmann, Eds., VCH Verlagsgessellschaft mbH, Federal Republic of Germany, **1**, 1996.

Wang, M.L. and H.M. Yang, *Ind. Eng. Chem. Res.*, **29**, 522, 1990.

Watanabe, Y. and T. Mukiyama, *Chem. Lett.*, No. 3, 285, 1981a.

Watanabe, Y. and T. Mukiyama, *Chem. Lett.*, No. 3, 443, 1981b.

Weber, W.P. and G.W. Gokel, *Phase-Transfer Catalysis in Organic Synthesis*, Springer-Verlag, New York, 1977.

Weissermel, K. and H.–J. Arpe, *Industrial Organic Chemistry*, 3rd Revised Edition, VCH Verlagsgesellschaft mbH, Weinheim, Federal Republic of Germany, 1997, p. 252.

Wrobel, J.T. and E. Hejchman, *Synthesis,* **5**, 452, 1987.

Chapter 4

ABSORPTION WITH REACTION

Jerry H. Meldon

4.1 Introduction

Design of absorbers and strippers requires accurate estimation of the effects of liquid-phase reaction upon interphase transfer *rates* and *selectivity* for targeted species. The reaction-related effect upon solvent *capacity* for a particular gas is naturally also important; the larger the capacity, the smaller the required solvent circulation rate. However, unless the reaction is sufficiently *fast*, a prohibitively tall column may be needed to exploit the capacity. Thus, *rates*—of both reaction and absorption—are the key. Although the focus here is upon use of first principles to estimate effects of chemical reaction upon absorption rates within *packed* towers, the concepts are applicable to other column figurations and to gas–liquid reactors (Carra and Morbidelli, 1987).

The sheer number of such rate estimates required to optimize absorber–stripper design motivates development of efficient algorithms. In each pass through a column, absorption rates must be calculated at numerous axial locations (z in Figure 4.1). Countercurrent flow of gas and liquid necessitates trial-and-error to determine the properties of effluent streams. Coupling of absorbers to strippers introduces yet another dimension of iteration. Finally, process optimization requires numerous such runs. Thus, despite the breathtaking speed of modern data processors, the development of accurate, computationally efficient algorithms remains desirable.

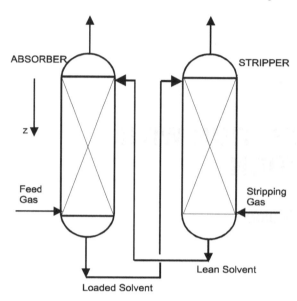

Figure 4.1:   Absorber–stripper system.

Quantitative description of transport in packed columns was formulated long ago in terms of gas- and liquid-phase mass transfer coefficients (k_G and k_{Lo}, respectively, where o here implies the absence of reaction effects). Procedures to estimate these and other design parameters including pressure drop and interfacial area per unit packed volume, a, are widely available (Danckwerts and Sharma, 1966; Charpentier, 1981; Zarzycki and Chacuk, 1993; Billet, 1995; Zenz, 1997; Seader and Henley, 1998). The definitive reference on gas scrubbing was recently updated (Kohl and Nielsen, 1997).

An abiding challenge, and our focus here, is the development of simple yet accurate algorithms for calculating the enhancement factors, E, by which reactions multiply absorption rates (Astarita, 1967; Danckwerts, 1970; Onda et al., 1970; Sherwood et al., 1975; Astarita et al., 1983; Doraiswamy and Sharma, 1984; Zarzycki and Chacuk, 1993).

4.2   Local Mass Transfer Models

Since early in the 20th century, mathematical models have been proposed to describe mass transfer between a gas phase and liquid flowing over and between column packing. The simple film model was proposed by

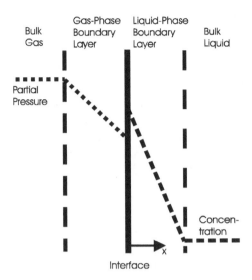

Figure 4.2:   Film theory conceptualization.

Whitman (1923) and first applied by Hatta (1928–1929) to absorption with chemical reaction.

As pictured in Figure 4.2, according to film theory, turbulence homogenizes bulk fluids within horizontal planes—except within the thin, unstirred fluid layers that prevail at phase interfaces because of the no-slip condition. Transport through these hypothetical films is assumed to proceed via steady-state molecular diffusion.

On the other hand, according to surface-renewal theory (Higbie, 1935; Danckwerts, 1951), elements of liquid are exposed to gas and transport proceeds for finite time intervals, θ, after which the elements are replaced instantaneously by well-mixed bulk liquid. The various surface-renewal models are distinguished by their assumed distributions of θ among fluid element populations. Higbie assigned a single value of θ; Danckwerts derived a distribution assuming that at any given instant, all fluid elements are equally likely to be replaced.

A distinguishing feature of analysis based on the steady-state film model is the implication that mass transfer coefficients are proportional to diffusion coefficients, D, whereas the transient surface-renewal models require proportionality to $\sqrt{D}$. Experimental data support the square-root dependence. Thus, corrections are necessary if one would like to exploit the film model's mathematical simplicity. Fortunately:

- Use of experimental k_{Lo} values in conjunction with E values computed from suitably corrected film model analyses yields excellent estimates of absorption rates.
- It is simple to correct film model analyses to satisfactorily estimate enhancement factors.

Olander (1960) showed that for cases of equilibrium (i.e., instantaneous, reversible) reaction and equal reactant diffusivities, identical expressions for E emerge from analyses based on the two different models. Chang and Rochelle (1982) showed, again for equilibrium reaction, that even with unequal D values, the two models' E values converge when D ratios appearing in film model analyses are replaced by ratios of the corresponding square roots. Later, Glasscock and Rochelle (1989) validated the square-root correction even for cases of nonequilibrium reaction, confirming the early, astute observations of Brian et al. (1961).

Subject to such correction, the film model suffices for process design. Before proceeding to illustrate its use, we derive an expression for the local absorption rate, N_A, *in the absence of reaction*, in terms of the bulk gas partial pressure, p_A^b, and bulk liquid concentration of transferred species A, $[A]^L$.

Steady-state rates of nonreactive species transport in gas and liquid are equal; thus,

$$N_A = k_G(p_A^b - p_A^o) = k_{Lo}([A]^o - [A]^L) \tag{4.1}$$

where *superscript o* here denotes the gas–liquid interface. Under most practical operating conditions, phase equilibrium may safely be assumed to prevail at the interface. We express this simply by

$$[A]^o = \alpha p_A^o \tag{4.2}$$

where the solubility coefficient α is a function of temperature and the natures of A and solvent.

From equations 4.1 and 4.2 it follows that

$$N_A = \frac{k_G([A]^* - [A]^L)}{\alpha + k_G/k_{Lo}} \tag{4.3}$$

where

$$[A]^* = \alpha p_A^b \tag{4.4}$$

It follows that transfer is liquid-phase controlled when $k_G/(k_{Lo}\alpha) \gg 1$ and gas-phase controlled when the sense of the inequality is reversed. Conservation of species A (referring to coordinate z in Figure 4.1) is expressed by

$$\frac{d}{dz}(u_L[A]^L) = -\frac{d}{dz}\left(\frac{u_G p_A^b}{RT}\right) = N_A a \qquad (4.5)$$

where u denotes superficial velocity (volumetric flow rate per unit column cross-section). Integration of equation 4.5 yields the requisite column height, H (Seader and Henley, 1998). We now can proceed to analyze the effects of reaction on N_A.

4.3 Mass Transfer with Chemical Reaction

All reactions, including the following between dissolved gas A and nonvolatile solute B, are reversible:

$$A + B \leftrightarrow C \qquad (4.6)$$

However, the choice of operating conditions determines the degree of reversibility.

Elevated pressure promotes absorption. Furthermore, absorbers are generally operated at lower temperatures, though not so low that either the rate of reaction or, if it is endothermic, its equilibrium constant, is severely compromised. The same principles, and generally opposite conclusions, guide selection of stripper conditions. In practice it is often the case that reactions are effectively irreversible under the conditions within a given column.

4.3.1 Irreversible Reaction

4.3.1.1 Film Model

When one applies the film model to an effectively irreversible reaction, as in equation 4.6, and if the diffusivities are essentially constant (e.g., in dilute solutions), the governing equations are

$$D_A \frac{d^2[A]}{dx^2} = D_B \frac{d^2[B]}{dx^2} = k_1[A][B] \qquad (4.7)$$

where k_1 is the reaction rate constant. The boundary conditions are defined by equation 4.1,

$$\frac{d[B]}{dx} = 0 \quad \text{at} \quad x = 0 \qquad (4.8)$$

which reflects the nonvolatility of B, and

$$[A] = [A]^L, [B] = [B]^L \quad \text{at} \quad x = L \qquad (4.9)$$

In addition, we retain the simple phase equilibrium expression, equation 4.2. Local mass balances, equation 4.5, are now expressed by

$$\frac{d}{dz}\left(\frac{u_G p_A^b}{RT}\right) = -N_A a \tag{4.10}$$

and

$$\frac{d}{dz}(u_L[i]^L) = -D_i a\left(\frac{d[i]}{dx}\right)_{x=L} - k_1[A]^L[B]^L h_L \qquad (i = A, B) \tag{4.11}$$

where h_L is liquid hold-up (vol./vol.) and the volume of the liquid film is assumed negligible. It follows from equation 4.11 that $[A]^L \to 0$ when

$$k_1[B]^L h_L \gg \frac{D_A a}{L} \quad \text{and} \quad \frac{u_L}{H} \tag{4.12}$$

i.e., when the time required for reaction is small compared to the respective times required for diffusion and passage through the column.

The nonlinear kinetic term precludes closed-form solution to equation 4.7 except in limiting cases such as the following.

When the reaction is effectively instantaneous, i.e., the Hatta number defined by

$$\mathrm{Ha} \equiv L\left(\frac{k_1[B]^L}{D_A}\right)^{1/2} \tag{4.13}$$

$\gg 1$, reactant concentrations vanish at some $x^\bullet < L$ (in which case $[A]^L$ *must* be 0). The location of $x^\bullet$ must be such that the fluxes of A from the interface and of B from bulk liquid are in stoichiometric ratio (1:1 here), i.e.,

$$\frac{D_A[A]^o}{x^\bullet} = \frac{D_B[B]^L}{L - x^\bullet} \tag{4.14}$$

It follows that

$$[A]^o = \frac{(\gamma - \beta_1)[A]^*}{1 + \gamma}, \quad N_A = \frac{D_A[A]^o + D_B[B]^L}{L}, \quad \text{and} \quad E = 1 + \beta_1 \tag{4.15}$$

where

$$\gamma \equiv \frac{k_G L}{D_A \alpha}, \quad \beta_1 \equiv \frac{D_B[B]^L}{D_A[A]^*} \tag{4.16}$$

and E is the enhancement factor by which reaction multiplies the absorption rate.

The above analysis requires that $N_A \le k_G p_A^b$ $(\beta_1 \le \gamma)$ so that in equation 4.15, $[A]^o > 0$; i.e., the absorption rate is upper-bounded by the maximum possible gas-phase flux.

When $\gamma \gg \beta_1$, $[A]^o \to [A]^*$. When in addition $\beta_1 \gg 1$, the absorption rate is limited only by the rate of diffusion of B to the interface.

Alternatively, if reaction is slow in the sense that $\mathrm{Ha}^2 \ll \beta_1$, $[B] \approx [B]^L$ everywhere. Reaction is then pseudo-first order, with $k \, (\equiv k_1[B]^L)$ replacing $k_1[B]$. Then, the solution to equation 4.7 is

$$[A] = \frac{[A]^o \sinh\left[\mathrm{Ha}\left(1 - \frac{x}{L}\right)\right] + [A]^L \sinh\left[\mathrm{Ha}\frac{x}{L}\right]}{\sinh \mathrm{Ha}} \qquad (4.17)$$

where

$$[A]^o = \frac{[A]^* + \dfrac{\mathrm{Ha}[A]^L}{\gamma \sinh \mathrm{Ha}}}{1 + \dfrac{\mathrm{Ha}}{\gamma \sinh \mathrm{Ha}}} \qquad (4.18)$$

As before, $[A]^o \to [A]^*$ when the gas-phase resistance is negligible, which in this case requires that

$$\gamma \gg \frac{\mathrm{Ha}}{\tanh \mathrm{Ha}} \qquad (4.19)$$

Then the absorption rate, in general, becomes

$$N_A = \frac{D_A \gamma \mathrm{Ha}}{L(\gamma \tanh \mathrm{Ha} + \mathrm{Ha})}\left([A]^* - \frac{[A]^L}{\cosh \mathrm{Ha}}\right) \qquad (4.20)$$

and so

$$E = \frac{(1 + \gamma)\mathrm{Ha}([A]^* - [A]^L / \cosh \mathrm{Ha})}{(\gamma \tanh \mathrm{Ha} + \mathrm{Ha})([A]^* - [A]^L)} \qquad (4.21)$$

When $[A]^L = 0$ and γ is large in the sense defined above, $E \to \mathrm{Ha}/\tanh \mathrm{Ha}$. When, furthermore, $\mathrm{Ha} > 3$, $E \to \mathrm{Ha}$ and $N_A \to \mathrm{Ha}D_A[A]^*$. Thus, absorption becomes independent of liquid film thickness because $[A] \to 0$ again at some $x^\bullet < L$. Note that when $\mathrm{Ha} > 3$, the pseudo-first-order assumption requires that $\beta_1 \gg 9$. It is also interesting to note (still regarding pseudo-first-order behavior) that absorption can occur even with $[A]^L > [A]^o$ (but *not* $> [A]^*$), since $\cosh \mathrm{Ha} \ge 1$.

4.3.1.2 Surface-Renewal Models

If one instead adopts a more realistic transient, i.e., the surface-renewal model, the mathematics is complicated by addition of time derivatives, e.g.,

$$D_i \frac{\partial^2 [i]}{\partial x^2} - \frac{\partial [i]}{\partial t} = k_1 [A][B] \qquad (i = A, B) \tag{4.22}$$

The boundary conditions at $x = 0$ remain unchanged. However, those previously applicable at $x = L$ are now enforced at $x = \infty$. For simplicity, we neglect gas-phase mass transfer resistance and therefore set $[A]^o = [A]^*$, and assume that initial conditions everywhere are $[A] = 0$ and $[B] = [B]^L$.

In the limiting case of instantaneous reaction, the solution (Danckwerts, 1970) is

$$E = \frac{1}{\mathrm{erf}\, \mu} \tag{4.23}$$

where μ is required to satisfy

$$\exp\left(\frac{\mu^2}{q}\right) \mathrm{erfc}\left(\frac{\mu}{\sqrt{q}}\right) = \frac{\beta_1}{\sqrt{q}} \exp(\mu^2)\, \mathrm{erf}\, \mu \tag{4.24}$$

and $q \equiv D_B / D_A$.

When $E \gg 1$, equations 4.23 and 4.24 reduce to

$$E = \frac{1 + \beta_1}{\sqrt{q}} \tag{4.25}$$

Whether or not $E \gg 1$, when $q - 1$ (i.e., diffusivities are equal), the expression for E is identical to that which emerged from the film model, equation 4.15.

Figure 4.3 compares enhancement factors for instantaneous forward, effectively irreversible, reaction (equation 4.6) based on surface-renewal theory. Equation 4.25 is indeed seen to be valid when $E \gg 1$. When $D_B = D_A$, equation 4.25 reduces to the expression for E based on film theory, equation 4.15. Furthermore, *in this case of equal diffusivities*, the latter simple expression yields E values identical to those represented by the middle solid line in Figure 4.3, i.e., the surface-renewal and film model analyses yield identical results.

Correspondingly, when the reaction is pseudo-first-order and $[A]^L = 0$, the model of Higbie (1935) gives

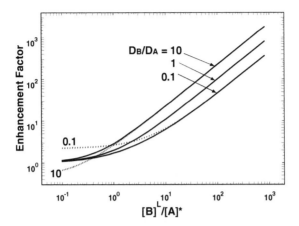

Figure 4.3: Enhancement factors for instantaneous irreversible surface-renewal theory. Solid lines: equations 4.23–4.24. Broken lines: equation 4.25.

$$E = \frac{\pi}{4\text{Ha}}\left[\left(\frac{4\text{Ha}^2}{\pi}+\frac{1}{2}\right)\text{erf}\left(2\text{Ha}/\sqrt{\pi}\right)+\frac{2\text{Ha}}{\pi}\exp\left(-4\text{Ha}^2/\pi\right)\right]$$

(4.26)

while that of Danckwerts (1951) yields

$$E = \sqrt{1+\text{Ha}^2}$$

(4.27)

In either case, when $\text{Ha} \gg 1$, $E \to \text{Ha}$, which again agrees with what emerges from the film model.

Figure 4.4 compares enhancement factors for first-order irreversible reaction $A \to$ products, based on what equation 4.21 (film model) reduces to when γ is large and $[A]^L = 0$ (squares), plus surface-renewal model equations 4.26 (circles) and 4.27 (solid curve), as well as $E = \text{Ha}$ (straight line), which yields the large Ha asymptote. Once again, enhancement factors based on the two models are nearly identical.

4.3.1.3 Van Krevelen/Hoftijzer (VKH) Analysis

Van Krevelen and Hoftijzer (1948) derived an approximate, yet remarkably accurate solution to equations 4.7. Before outlining it, we transform to the following dimensionless variables for conciseness:

$$A \equiv \frac{[A]}{[A]^*} \qquad B \equiv \frac{[B]}{[B]^L} \qquad y \equiv \frac{x}{L}$$

(4.28)

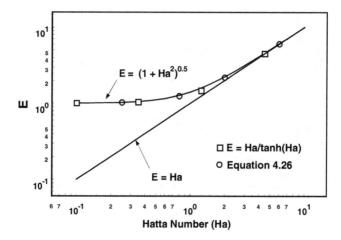

Figure 4.4: Comparison of theoretical enhancement factors for first-order irreversible reaction.

Equations 4.7 then become

$$\frac{\mathrm{d}^2 A}{\mathrm{d}y^2} = \beta_1 \frac{\mathrm{d}^2 B}{\mathrm{d}y^2} = \mathrm{Ha}^2 AB \tag{4.29}$$

subject to

$$\frac{\mathrm{d}A}{\mathrm{d}y} = -\gamma(1 - A), \quad \frac{\mathrm{d}B}{\mathrm{d}y} = 0 \quad \text{at} \quad y = 0 \tag{4.30}$$

$$A = 0, \quad B = 1 \quad \text{at} \quad y = 1 \tag{4.31}$$

The derivative terms in equation 4.29 may be subtracted from one another to eliminate the reaction term. When the difference is integrated twice, the result is

$$A - \beta_1 B = -\phi y + \eta \tag{4.32}$$

where ϕ $(= (\mathrm{d}A/\mathrm{d}y)_{y=0})$ and η are constants; ϕ may be identified as the "dimensionless absorption rate."

It follows from the boundary condition $(\mathrm{d}B/\mathrm{d}y)_{y=0} = 0$ that $B \approx B^o$ in the vicinity of $y = 0$. In view of this and the particular importance of the boundary region for determining the absorption rate, van Krevelen and Hoftijzer linearized the first of equations 4.29 as follows:

$$\frac{\mathrm{d}^2 A}{\mathrm{d}y^2} \approx \rho_1^2 A \tag{4.33}$$

where $\rho_1 \equiv (HaB^o)^{1/2}$ and B^o remains to be determined. Note that this is *not* the same approximation as that made in the case of a slow, pseudo-first-order reaction, i.e., $[B] \approx [B]^L$.

The solution to equation 4.33, subject to the boundary conditions for A, is

$$A \approx \frac{\sinh[\rho_1(1-y)]}{\sinh \rho_1 + (\rho_1/\gamma)\cosh \rho_1} \qquad (4.34)$$

Inserting the boundary conditions for both A and B in equation 4.32 and combining the result with equation 4.34 lead to

$$\phi = A^o + \beta_1(1 - B^o) = \gamma(1 - A^o) = \frac{\rho_1}{\tanh \rho_1 + \rho_1/\gamma} \qquad (4.35)$$

which together determine A^o, B^o, and ϕ.

Hikita and Asai (1964) extended the VKH linearization to accommodate the generalized reaction rate $k[A]^m[B]^n$ (with m and n integers); see also Asai (1990). DeCoursey (1974, 1982) and DeCoursey and Thring (1989) applied it within the context of surface-renewal theory.

We proceed to absorption with *reversible* reaction, for which irreversible reaction is simply a limiting case.

4.3.2 Reversible Reaction

Peaceman (1951) extended the VKH approach to cases of reversible reaction. For reaction 4.6, the governing equations are

$$\frac{d^2A}{dy^2} = \beta_1 \frac{d^2B}{dy^2} = -\beta_1\beta_2 \frac{d^2C}{dy^2} = Ha^2\left(AB - \frac{C}{K}\right) \qquad (4.36)$$

where

$$\beta_2 \equiv \frac{D_C}{D_B} \qquad K \equiv \frac{k_1[A]^*}{k_2} \qquad C = \frac{[C]}{[B]^L} \qquad (4.37)$$

k_2 is the reverse rate constant and K the dimensionless equilibrium constant.

Combining the derivative expressions for A and C, and again integrating twice lead to

$$A + \beta_1\beta_2 C = -\phi y + \omega \qquad (4.38)$$

where ω is a constant.

The presence of sufficient liquid hold-up (which ensured that $[A]^L = 0$ in the case of irreversible reaction) now implies *equilibrium* in bulk liquid, i.e.,

$$C^L = KA^L \qquad (4.39)$$

Combining equations 4.38 and 4.39 with equation 4.36 applied to A, and again fixing B at B^o, yields

$$\frac{d^2 A}{dy^2} = \rho_2^2 A + \frac{\text{Ha}^2(\phi y - \omega)}{K\beta_1\beta_2} \tag{4.40}$$

where

$$\rho_2 \equiv \text{Ha}\sqrt{B^o\left(1 + \frac{1}{K\beta_1\beta_2}\right)} \tag{4.41}$$

It follows from the solution to equation 4.40 allowing for gas-phase mass transfer resistance, that

$$\phi = \gamma(1 - A^o) = A^o - A^L + \beta_1(1 - B^o)$$

$$= \frac{1 + K\beta_1\beta_2 B^o - A^L\left[\dfrac{K\beta_1\beta_2(B^o - 1)}{\cosh\rho_2} - 1 - K\beta_1\beta_2\right]}{(1 + K\beta_1\beta_2)/\gamma + K\beta_1\beta_2 B^o \tanh[\rho_2/(\rho_2 - 1)]} \tag{4.42}$$

Equations 4.42 determine A^o, B^o, and ϕ.

As noted below, VKH linearization introduces surprisingly small error in enhancement factors. Nonetheless, the open literature contains no example of its successful application to absorption with more complex chemical kinetics of practical interest. This motivated the work described in the following section.

4.3.3 Perturbation Methods

Meldon et al. (1995) employed an alternative approach to solving equations 4.36 based on the singular perturbation methods Smith et al. (1973) had applied to analyze diffusion and reversible reaction in membranes. Its extension to complex reaction schemes is straightforward, as illustrated further below in a practical example. The key to this approach is the breakdown of local concentrations into equilibrium and perturbation components: in dimensionless terms $(\mathbf{A}, \mathbf{B}, \mathbf{C})$ and $(\delta A, \delta B, \delta C)$, respectively, as defined by

$$A = \mathbf{A} + \delta A, \quad B = \mathbf{B} + \delta B, \quad \text{and} \quad C = \mathbf{C} + \delta C \tag{4.43}$$

$$\mathbf{A} + \beta_1\mathbf{B} = -\phi y + \eta \tag{4.44}$$

$$\mathbf{A} + \beta_1\beta_2\mathbf{C} = -\phi y + \omega \tag{4.45}$$

$$\frac{\mathbf{C}}{\mathbf{A}\mathbf{B}} = K \tag{4.46}$$

It follows that

$$\frac{d^2\mathbf{A}}{dy^2} + \frac{d^2}{dy^2}(\delta A) = \text{Ha}^2 f \delta A + O(\delta A^2) \tag{4.47}$$

where

$$f \equiv \frac{\mathbf{A}}{\beta_1} + \mathbf{B} + \beta_2 \mathbf{C} \tag{4.48}$$

Retaining only terms first order in δA, equation 4.47 reduces to

$$\frac{d^2}{dy^2}(\delta A) = \text{Ha}^2 f \delta A \tag{4.49}$$

and so

$$\delta A = \delta A^o \frac{\sinh[\text{Ha}\sqrt{f}(1-y)]}{\sinh[\text{Ha}\sqrt{f}]} \tag{4.50}$$

The unknowns δA^o, $\mathbf{A}^o$, ϕ, and thus enhancement factor E, are determined by the boundary conditions.

Figure 4.5 compares E values calculated for a case in which $[A]^L = 0$, the diffusion coefficients are all equal, and gas-phase mass transfer resistance is negligible. Results are based on (a) the two approximate solutions to equation 4.36 described above and (b) DeCoursey's (1982) application of VKH linearization to the partial differential equations that evolve in the context of

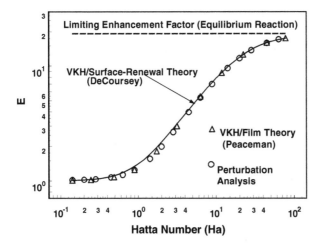

Figure 4.5: Comparison of theoretical enhancement factors $(A + B = C$ reversible).

surface-renewal theory. Note that for purposes of comparison, surface-renewal theory's characteristic renewal time, θ, is equated with film theory's characteristic diffusion time, L^2/D.

The results in Figure 4.5 illustrate the excellent agreement among E values derived from solutions based on VKH approximation (DeCoursey's using the surface-renewal model, and Peaceman's using the film model), and that based on singular perturbation methods using the film model. Notably, DeCoursey's analysis has been shown to yield E values in nearly exact agreement with those derived via numerical analysis of the same differential equations. This reaffirms the reliability of film theory in the case of equal diffusivities. A *high*-Ha asymptote is approached when the reaction is sufficiently fast that chemical equilibrium effectively prevails locally throughout the liquid film.

Figure 4.6 depicts enhancement factors calculated for the reaction $A + B = C + D$, again with gas-phase mass transfer resistance neglected and $[A]^L = 0$, but this time with unequal diffusion coefficients ($[B]^L/[A]^* = 100$, $K = 1$).

Enhancement factors based on unmodified film theory using the singular perturbation analysis disagree with those derived from DeCoursey's application of VKH linearization to surface-renewal theory which, as noted already, has been independently validated by comparison of the E values it predicts with those based on exact numerical methods. It is, therefore, significant that when diffusivity ratios (D_i/D_A; $i = B, C, D$) are replaced

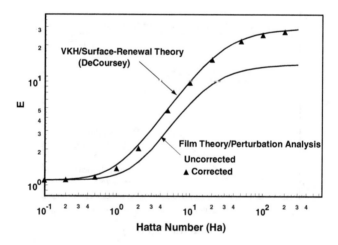

Figure 4.6: Comparison of theoretical enhancement factors for $A + B = C + D$ reversible. ($D_i/D_A = 0.2$; $i = B, C, D$.)

with square roots of the same ratios, the agreement of results from analyses based on film and surface-renewal theories is restored.

Having validated singular perturbation analysis and the corrected film theory in the case of simple reactions, we apply a similar analysis in the next section to a more complex problem of practical interest. The reader is referred to Table 4.1 for a list of recent publications of related interest.

4.4 Simultaneous Absorption of Carbon Dioxide and Hydrogen Sulfide in Alkaline Solutions

When an aqueous solution is exposed to a mixture containing the "acid gases" CO_2 and H_2S, the following reactions occur:

$$CO_2 + H_2O \leftrightarrow HCO_3^- + H^+ \tag{4.51}$$

$$CO_2 + OH^- \leftrightarrow HCO_3^- \tag{4.52}$$

$$HCO_3^- \leftrightarrow CO_3^{-2} + H^+ \tag{4.53}$$

$$H_2S \leftrightarrow HS^- + H^+ \tag{4.54}$$

$$H_2O \leftrightarrow H^+ + OH^- \tag{4.55}$$

The presence of alkali, typically sodium or potassium carbonate, ensures a sufficiently high pH that the extents of these reactions are appreciable. Significantly, reactions 4.53–4.55 are generally effectively instantaneous, while reactions 4.51 and 4.52 are slow. Because of this, film thickness L plays an important role in determining absorption selectivity. This is of great practical significance since it is often desirable to selectively remove H_2S from gases containing CO_2, e.g., in natural gas processing and whenever H_2S is to be converted to elemental sulfur in a downstream reactor, the cost of which increases with gas volume.

Dissolved gases diffuse more rapidly in thin films. Accordingly, it is only in sufficiently *thick* liquid films that diffusion is sufficiently slow for reactions 4.51 and 4.52 to enhance CO_2 transport. On the other hand, the instantaneous dissociation of dissolved H_2S *always* enhances its transport. Consequently, selectivity for H_2S is promoted by minimizing liquid film thickness. Notably, this is a rare instance in which *selectivity* and *rate* of absorption go hand in hand. Of course, promotion of H_2S is insignificant when liquid-phase transfer is so rapid that nonselective gas-phase mass transfer controls absorption.

Finally, when the liquid film is sufficiently thick, CO_2 absorption is enhanced by its otherwise "slow" reactions, and selectivity for H_2S again

Table 4.1: Recent Publications of Interest.

Reference	System	Focus
Aroonwilas and Tontiwachwuthikul (1998)	CO_2 in aq. AMP[a]	Structured vs. random packing
Ashour et al. (1996)	Cl_2 in aq. hydroxide solutions	Absorption kinetics
Aurousseau et al. (1996)	SO_2 in aq. Ce(IV)/acid	Absorption kinetics
Bhattacharya et al. (1996)	SO_2 in aq. dimethyl aniline dispersion	Absorption rate
Bravo et al. (1996)	$SO_2 + O_2$ in aq. $MnSO_4$	Absorption kinetics
Buzek and Jaschik (1995)	SO_2 in aq. $NaHSO_3/Na_2SO_3/Na_2SO_4$	Equilibria
Camacho et al. (1995)	O_2 in aq. NaS_2O_4	pH-dependence of absorption rate
Chaudhuri et al. (1997)	CO in aq. NaOH/hexane dispersions	Absorption/phase transfer catalysis
Dawodu and Meisen (1996a)	CO_2 in aq. alkanolamine blends	Amine degradation
Dawodu and Meisen (1996b)	CS_2 in aq. DEA^b	Amine degradation
Demmink et al. (1997)	NO in aq. Fe chelate solution	Absorption kinetics
Frank et al. (1996)	NH_3 in water	Multicomponent diffusion
Gerard et al. (1996)	SO_2 in aq. $CaSO_3$ slurries	Model
Griolet et al. (1996)	Phosgene in aq. NaOH	Absorption kinetics
Hagewiesche et al. (1995)	CO_2 in aq. $MEA^c/MDEA^d$	Absorption kinetics
Haji-Sulaiman and Aroua (1996)	CO_2 in aq. DEA^b and AMP^a	Equilibria
Haji-Sulaiman et al. (1996)	CO_2 in aq. alkanolamine solutions	Equilibria
Hofele et al. (1996)	NO in aq. Fe chelate solution	Equilibria
Hogendoorn et al. (1995)	CO in aq. Cu salt solution	Absorption rate and equilibria
Lammers et al. (1995)	CO_2, H_2S, COS in aq. $MDEA^d$/polyhydroxyalcohols	Absorption rates
Li et al. (1998)	H_2S in aq. NaOH	Hollow fiber membranes
Mathonat et al. (1998)	CO_2 in aq. MEA^c	Calorimetry
Pak and Suzuki (1997)	CO_2 in aq. alkanolamine	Power plant efficiency
Pradhan et al. (1997)	NOx in aq. HNO_3	Model

Reference	System	Topic
Rascol et al. (1996)	CO_2, H_2S in aq. mixed alkanolamine solutions	Simulation
Rho et al. (1997)	CO_2 in aq. MDEA[d]	Equilibria
Riazi (1996)	Absorption/reaction in turbulent films	Model
Rinaldi (1997)	CO_2, H_2S, SO_2 in aq. polyalkenepolyaminephenols	Absorption and desorption
Rinker et al. (1995)	CO_2 in aq. MDEA[d]	Absorption kinetics
Rinker et al. (1996)	CO_2 in aq. DEA[b]	Absorption kinetics
Rogers et al. (1997)	CO_2, H_2S in aq. DEA[b]	Equilibria
Safarik and Eldridge (1998)	Olefins, paraffins in Cu and Ag complexing solutions	Review
Saha et al. (1995)	CO_2 in aq. AMP[a]	Absorption kinetics
Sanders et al. (1995)	SO_2 in fly ash/hydrated lime slurries	Absorption in a spray drier
Shaikh et al. (1995)	Bubbles and films	Theory
Shi et al. (1996)	NO in aq. Fe chelate solution	Absorption kinetics
Suda et al. (1997)	CO_2 in aq. alkanolamines	Absorption from flue gas
Taniguchi et al. (1997)	CO_2 in aq. NaOH	Spray column
Thomas and Vanderschuren (1996)	NO in aq. H_2O_2	Absorption/oxidation
Thomas and Venderschuren (1997)	NOx in aq. HNO_3/H_2O_2	Model
Thomas and Venderschuren (1998)	NOx in aq. HNO_3/H_2O_2	Model/temperature dependence
Vas Bhat et al. (1997)	Heat effects in absorption with reversible reaction	Analysis
Vilcu et al. (1997)	CO_2 in diisopropanol amine/PEGDME[e]	Equilibria
Xu et al. (1995)	CO_2 in aq. MDEA[d]/piperazine	Desorption rate
Xu et al. (1996)	CO_2 in aq. AMP[a]	Absorption kinetics

[a] AMP = 2-amino-2-methyl-1-propanol.
[b] DEA = diethanol amine.
[c] MEA = monoethanol amine.
[d] MDEA = methyldiethanol amine.
[e] PEGDME = polyethyleneglycol dimethyl ether.

vanishes. In order to identify operating conditions that maximize selectivity for H_2S, it is necessary to solve the corresponding differential equations for this more complex reactive scheme.

The singular perturbation method outlined above for a case of simple chemical reaction is directly applicable here (Al-Hashimi, 2000) (see this thesis for validation of the analysis by comparison of its predictions with results of exact numerical analysis). Characteristic values for the enhancement factors of CO_2 and H_2S, and selectivity for H_2S (defined as ratio of enhancement factors times ratio of physical solubilities) including typical gas-phase mass transfer effects, are depicted in Figure 4.7. These results suggest the possibility of extremely high selectivity for H_2S. Selectivity for H_2S vanishes at both high and low film thicknesses: (a) for low film thicknesses, because gas-phase mass transfer, which is nonselective, controls absorption rates; (b) for high film thicknesses, because of reaction-enhanced CO_2 absorption when diffusion is very slow (compared to even the slow reactions of CO_2) in sufficiently thick films. Notably, the selectivity maximizes over a range of liquid film thicknesses, 10^{-3} to 10^{-2} cm, characteristic of turbulent flow in packed columns. Thus, the basis of highly selective H_2S absorption is clear.

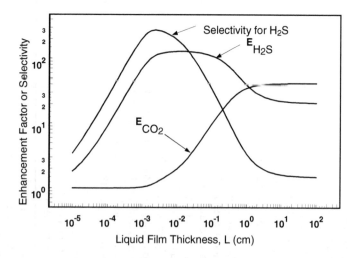

Figure 4.7:   Theoretical E values for CO_2 and H_2S, and selectivity in 2M K_2CO_3.

4.5 Conclusions

It is apparent, then, that approximate analytical methods, which typically require an order of magnitude less computer time than numerical methods, are sufficiently accurate to warrant deployment in process design. Furthermore, because they promote awareness of underlying physicochemical phenomena, such analyses are attractive not only as adjuncts to design software, but as educational tools.

Symbols

a	Interfacial area per unit volume of column, $m^2\,m^{-3}$
A, B	Defined by equation 4.28
$\delta A, \delta B, \delta C$	Defined by equation 4.43
$\mathbf{A}, \mathbf{B}, \mathbf{C}$	Defined by equations 4.44–4.46
C	Defined by equation 4.37
D	Diffusion coefficient, $m^2\,s^{-1}$
E	Enhancement factor; ratio of absorption rates with and without reaction
f	Defined by equation 4.48
h_L	Liquid hold-up, $m^3\,m^{-3}$
H	Column height, m
Ha	Hatta number; defined by equation 4.13 for the case of second-order reaction
k_G	Gas-phase mass transfer coefficient, $mol\,atm^{-1}\,m^{-2}\,s^{-1}$
k_L	Liquid-phase mass transfer coefficient, $m\,s^{-1}$
k_1	Second-order forward reaction rate constant, $m^3\,mol^{-1}\,s^{-1}$
k_2	First-order reverse reaction rate constant, s^{-1}
K	Equilibrium constant of reaction $A + B = C + D$; or as defined in equation 4.37
L	Liquid film thickness, m
N	Species flux, $mol\,m^{-2}\,s^{-1}$
q	D_B/D_A
t	Time, s
u	Superficial velocity, $m^3\,m^{-2}\,s^{-1}$
x	Coordinate as defined in Figure 4.2, m
$x^{\bullet}$	Defined by equation 4.14
y	Defined by equation 4.28
z	Coordinate as defined in Figure 4.1, m

Greek

α	Solubility coefficient, mol m^{-3} atm^{-1}
β_1	Defined by equation 4.16
β_2	Defined by equation 4.37
γ	Defined by equation 4.16
η	Defined by equation 4.32
μ	Defined by equation 4.24
ρ_1	$(\mathrm{Ha}B^o)^{1/2}$
ρ_2	Defined by equation 4.41
ϕ	Defined by equation 4.32
ω	Defined by equation 4.38

Subscripts and Superscripts

A, B, C, D	Species A, B, C, D
b	Bulk gas
G	Gas phase
L	Liquid phase, or condition at $x = L$
o	Gas–liquid interface, or condition at $x = 0$
$*$	Equilibrium with bulk gas

References

Al-Hashimi, S., Ph.D. Thesis, Tufts University, Medford, MA, 2000.

Aroonwilas, A. and P. Tontiwachwuthikul, *Ind. Eng. Chem. Res.*, **37**, 569, 1998.

Asai, H., *Can. J. Chem. Eng.*, **68**, 284, 1990.

Ashour, S.S., E.B. Rinker, and O.C. Sandall, *AIChE J.*, **42**, 671, 1996.

Astarita, G., *Mass Transfer with Chemical Reaction*, Elsevier, Amsterdam, 1967.

Astarita, G., D.W. Savage, and A. Bisio, *Gas Treating with Chemical Solvents,* Wiley & Sons, New York, 1983.

Aurousseau, M., C. Roizard, A. Storck, and A. Lapicque, *Ind. Eng. Chem. Res.,* **35**, 1243, 1996.

Bhattacharya, S., B.K. Dutta, M. Shyamal, and R.K. Basu, *Can. J. Chem. Eng.,* **74**, 339, 1996.

Billet, R., *Packed Towers,* VCH Publishers, New York, 1995.

Bravo, V.R., F.R. Camacho, and M.V. Moya, *Can. J. Chem. Eng.,* **74**, 104, 1996.

Brian, P.L.T., J.F. Hurley, and E.H. Hasseltine, *AIChE J.*, **7**, 226, 1961.

Buzek, J. and M. Jaschik, *Chem. Eng. Sci.*, **50**, 3067, 1995.

Camacho, F., M.P. Paez, G. Blazquez, M.A. Jimenez, and M. Fernandez, *Chem. Eng. Sci.*, **50**, 1181, 1995.

Carra, S. and M. Morbidelli, "Gas–Liquid Reactors," in *Chemical Reaction and Reactor Engineering*, J.J. Carberry and A. Varma, Eds., Marcel Dekker, New York, 1987.

Chang, C.-S. and G.T. Rochelle, *Ind. Eng. Chem. Fundam.*, **21**, 379, 1982.

Charpentier, J.C., *Adv. Chem. Eng.*, **11**, 1, 1981.

Chaudhari, R.V., P. Jayasree, S.P. Gupte, and H. Delmas, *Chem. Eng. Sci.*, **52**, 4197, 1997.

Danckwerts, P.V., *Ind. Eng. Chem.*, **4**, 1460, 1951.

Danckwerts, P.V., *Gas–Liquid Reactions*, McGraw-Hill, New York, 1970.

Danckwerts, P.V. and M.M. Sharma, *Chemical Engineer*, CE244, 1966.

Dawodu, O.F. and A. Meisen, *Can. J. Chem. Eng.*, **74**, 960, 1996a.

Dawodu, O.F. and A. Meisen, *Gas Sep. Purif.*, **10**, 1, 1996b.

DeCoursey, W.J., *Chem. Eng. Sci.*, **29**, 1867, 1974.

DeCoursey, W.J., *Chem. Eng. Sci.*, **37**, 1483, 1982.

DeCoursey, W.J. and R.W. Thring, *Chem. Eng. Sci.*, **44**, 1715, 1989.

Demmink, J.F., I.C.F. van Gils, and A.A.C.M. Beenackers, *Ind. Eng. Chem. Res.*, **36**, 4914, 1997.

Doraiswamy, L.K. and M.M. Sharma, *Heterogeneous Reactions*, Wiley & Sons, New York, **2**, 1984.

Frank, M.J.W., J.A.M. Kuipers, and W.P.M. van Swaaij, *Chem. Eng. Sci.*, **51**, 2619, 1996.

Gerard, P., G. Sagantini, and J. Vanderschuren, *Chem. Eng. Sci.*, **51**, 3349, 1996.

Glasscock, D.A. and G.T. Rochelle, *AIChE J.*, **35**, 1271, 1989.

Griolet, F., J. Lieto, and G. Astarita, *Chem. Eng. Sci.*, **51**, 3213, 1996.

Hagewiesche, D.P., S. Ashour, H. Al-Ghawas, and O.C. Sandall, *Chem. Eng. Sci.*, **50**, 1071, 1995.

Haji-Sulaiman, M.Z. and M.K. Aroua, *Chem. Eng. Commun.*, **140**, 157, 1996.

Haji-Sulaiman, M.Z., M.K. Aroua, and M.I. Pervez, *Gas Sep. Purif.*, **10**, 13, 1996.

Hatta, S., *Technol. Rep. Tohuku Imper. Univ.*, **8**, 1, 1928–29.

Higbie, R., *Trans. AIChE J.*, **35**, 365, 1935.

Hikita, H. and S. Asai, *Int. Chem. Eng.*, **4**, 332, 1964.

Hofele, J., D. van Velzen, H. Langenkamp, and K. Schaber, *Chem. Eng. Process.*, **35**, 295, 1996.

Hogendoorn, J.A., W.P.M. van Swaaij, and G.F. Versteeg, *Chem. Eng. J.*, **59**, 243, 1995.

Kohl, A.L. and R.B. Nielsen, *Gas Purification*, 5th Edition, Gulf, Houston, 1997.

Lammers, J.N.J.J., J. Haringa, and R.J. Littel, *Chem. Eng. J.*, **60**, 123, 1995.

Li, K., D. Wang, C.C. Koe, and W.K. Teo, *Chem. Eng. Sci.*, **53**, 1111, 1998.

Mathonat, C., V. Majer, A.E. Mather, and J.-P.E. Grolier, *Ind. Eng. Chem. Res.*, **37**, 4136, 1998.

Meldon, J.H., M. Osias, and L. Olawoyin, Preprint 55b, AIChE Spring Mtg., Houston, 1995.

Olander, D.R., *AIChE J.*, **6**, 233, 1960.

Onda, K., E. Sada, T. Kobayashi, and M. Fujine, *Chem. Eng. Sci.*, **25**, 753, 1970.

Pak, P.S. and Y. Suzuki, *Int. J. Energy Res.*, **21**, 749, 1997.

Peaceman, D.W., Sc. D. Thesis, Mass. Inst. Tech., Cambridge, MA, 1951.

Pradhan, M.P., N.J. Suchak, P.R. Walse, and J.B. Joshi, *Chem. Eng. Sci.*, **52**, 4569, 1997.

Rascol, E., M. Meyer, and M. Prevost, *Comput. Chem. Eng.*, **20** (Suppl. B), S1401, 1996.

Rho, S.-W., K.-P. Yoo, J.S. Lee, S.C. Nam, J.E. Son, and B.-M. Min, *J. Chem. Eng. Data*, **42**, 1161, 1997.

Riazi, M.R., *Gas Sep. Purif.*, **10**, 41, 1996.

Rinaldi, G., *Ind. Eng. Chem. Res.*, **36**, 3778, 1997.

Rinker, E.B., S.S. Ashour, and O.C. Sandall, *Chem. Eng. Sci.*, **50**, 755, 1995.

Rinker, E.B., S.S. Ashour, and O.C. Sandall, *Ind. Eng. Chem. Res.*, **35**, 1107, 1996.

Rogers, W.J., J. Bullin, R. Davison, R.E. Frazier, and K.N. Marsh, *AIChE J.*, **43**, 3223, 1997.

Safarik, D.J. and R.B. Eldridge, *Ind. Eng. Chem. Res.*, **37**, 2571, 1998.

Saha, A.K., S.S. Bandyopadhyay, and A.K. Biswas, *Chem. Eng. Sci.*, **50**, 3587, 1995.

Sanders, J.F., T.C. Keener, and J. Want, *Ind. Eng. Chem. Res.*, **34**, 302, 1995.

Seader, J.D. and E.R. Henley, *Separation Process Principles,* Wiley & Sons, New York, 1998.

Shaikh, A.A., A. Jamal, and S.M. Zarook, *Chem. Eng. J.*, **57**, 27, 1995.

Sherwood, T.K., R.L. Pigford, and C.R. Wilke, *Mass Transfer*, McGraw-Hill, New York, 1975.

Shi, Y., D. Littlejohn, and S.G. Chang, *Ind. Eng. Chem. Res.*, **35**, 1668, 1996.

Smith, K.A., J.H. Meldon, and C.K. Colton, *AIChE J.*, **19**, 102, 1973.

Suda, T., M. Iijima, H. Tanaka, S. Mitsuoka, and T. Iwaki, *Environ. Prog.*, **16**, 200, 1997.

Taniguchi, I., Y. Takamura, and K. Asano, *J. Chem. Eng. Japan*, **30**, 427, 1997.

Thomas, D. and J. Vanderschuren, *Chem. Eng. Sci.*, **51**, 2649, 1996.

Thomas, D. and J. Vanderschuren, *Ind. Eng. Chem. Res.*, **36**, 3315, 1997.

Thomas, D. and J. Vanderschuren, *Ind. Eng. Chem. Res.*, **37**, 4418, 1998.

van Krevelen, D.W. and P.J. Hoftijzer, *Recl. Trav. Chim.*, **67**, 563, 1948.

Vas Bhat, R.D., W.P.M. van Swaaij, N.E. Benes, J.A.M. Kuipers, and G.F. Versteeg, *Chem. Eng. Sci.*, **52**, 4079, 1997.

Vilcu, R., I. Gainar, and G. Anitescu, *Rev. Roum. Chim.*, **42**, 63, 1997.

Whitman, W.G., *Chem. Met. Eng.*, **29**, 147, 1923.

Xu, G.-W., C.-F. Zhang, S.-J. Qin, and B.-C. Zhu, *Ind. Eng. Chem. Res.*, **34**, 874, 1995.

Xu, S., Y.-W. Wang, F.D. Otto, and A.E. Mather, *Chem. Eng. Sci.*, **51**, 841, 1996.

Zarzycki, R. and A. Chacuk, *Absorption*, Pergamon, Oxford, 1993.

Zenz, F.A., Section 3.2 in *Handbook of Separation Techniques for Chemical Engineers*, 3rd Edition, P.A. Schweitzer, Ed., McGraw-Hill, New York, 1997.

Chapter 5

ADSORPTION WITH REACTION

Robert W. Carr and Hemant W. Dandekar

5.1 Introduction

Reactive separation processes combine the normally separate and sequential unit operations of reaction and separation into a single, simultaneous operation. This coupling can lead to considerable savings in capital and operating costs. It can also help separate products which otherwise could not be separated, for example, those that form azeotropes. Reactive separation processes can also overcome equilibrium conversion limitations and sometimes enhance selectivity. These benefits provide a significant incentive to understanding and applying reactive separation processes in practical applications.

Reactive separation processes can be segregated into a number of classes depending on how the separation is carried out: whether it is through distillation, extraction, absorption, adsorption, or crystallization. This chapter will concentrate on adsorption with reaction. An adsorption-with-reaction process is defined as one in which reaction and separation using adsorption are accomplished simultaneously in a single unit operation.

Reactive chromatography as a means of effecting simultaneous reaction with adsorption has been known for close to 40 years (Basset and Habgood, 1960). The main drawback of this type of process for industrial applications is that the process operates discontinuously and a large volume of adsorbent and catalyst is required. Thus for an adsorption with reaction process to be commercially successful it needs a contacting scheme that allows continuous operation. A number of advantages can be realized if a continuous process replaces a conventional process that involves sequential adsorption and reaction steps.

The advantages are:

- Conversion beyond equilibrium can be achieved.
- Capital costs can be lower because of lower separation costs downstream. Also, a lower-severity operation, such as operation at lower pressures, can result in lower capital costs.
- Feed recycle can be reduced or eliminated, resulting in lower operating costs.
- Increase in selectivities of intermediate products can improve productivity or product quality or both.
- Some processes can be run at less severe conditions, resulting in longer catalyst life.

5.1.1 Applicability

The adsorption with reaction methodology cannot be applied to all processes. The processes that are most suitable involve:

- Equilibrium- or selectivity-limited reactions.
- Reactions in which products can be separated by adsorption.
- Compatible reaction and adsorption conditions. Often a number of vapor-phase reactions occur at high temperatures where little adsorption capacity is available. This requires design modifications in order to carry out reaction at high temperature and separation at a lower temperature.
- Use of adsorption with reaction process is applicable only where other separation processes such as extraction or distillation do not work. Usually methods such as extraction or distillation are simpler to implement than adsorption. For example, reaction with adsorption may be preferable when azeotropes or close-boiling products are present.

5.1.2 Attributes

The adsorption-with-reaction processes have the following attributes:

- *One-fluid phase.* The processes operate in one-fluid phase, either liquid or vapor. A second phase is the solid, where adsorption is carried out. The presence of more than one-fluid phase adversely affects mass transfer and may increase dispersion. Fast mass transfer and low dispersion are desirable characteristics for good separation.

- *Heterogeneous or homogeneous catalyst.* The catalyst can be either a solid or fluid. If a solid, it could be on the same particle as the adsorbent, or the catalyst and adsorbent could be separate particles in a physical admixture.
- *Compatibility of adsorption and reaction conditions.* This challenge is the biggest in developing adsorption with reaction processes. Often the temperature of the gas-phase reactions is too high, resulting in poor adsorbent capacity or selectivity. Physisorption is limited to a maximum temperature of 250–300°C. In liquid-phase reactions, mass transfer can be too slow for effective adsorptive separation, especially when a single product is formed.
- *Compatibility of desorbent or purge medium.* Understanding the effect of the desorbent on the catalyst and the effect of catalyst on the desorbent is extremely important. The desorbent may deactivate the catalyst, or the catalyst may cause the desorbent to react and lose its desirable adsorption characteristics or catalyze formation of harmful byproducts that may adversely affect the process.

5.2 Reactor Types

It has been known since the late 1950s that chemical reactions can be carried out in chromatographic packed columns capable of separating the reactive components. This comprises a class of separative chemical reactors in which the separation is effected by adsorption, and which has been termed reaction chromatography. Either homogeneous reactions or heterogeneous reactions may occur, the latter requiring a solid catalyst mixed with the chromatographic packing material. In both cases, reaction and separation occur in concert in a locally integrated process. The separation can significantly reduce the requirements for downstream processing. Furthermore, the separation of reactants and products can shift unfavorable chemical equilibria in the direction of reaction products, enhancing conversions and product yields.

For example, in a reaction of the type $A = B + C$, the separation of B from C suppresses the reverse reaction and the conversion of A can proceed beyond the equilibrium value that would be the maximum conversion that would be obtained in a nonseparative reactor. Furthermore, the approach to chemical equilibrium is always an approximately exponential decay, no matter what the form of the kinetic rate expression (Aris, 1969). The asymptotic approach to equilibrium requires either a long tubular flow reactor, a large-volume CSTR (continuously stirred tank reactor), or long contact time in a batch reactor to realize conversions that are close to chemical equilibrium, and are the maximum attainable in the absence of separation. A chromatographic reactor is capable of giving higher conversions, frequently approaching unity, in smaller reactors or in shorter reaction times, than nonseparative reactors. In spite of these advantages, conventional column reaction chromatography has not been commercialized. It is a batch process in which reactant is injected into the column at intervals, the processing time depending upon the retention times of the reactive components.

Continuous flow processes for reaction chromatography can be realized by introducing motion between a feed stream and the adsorbent/catalyst bed. Three configurations for accomplishing this are the rotating cylindrical annulus, the countercurrent moving bed, and the simulated countercurrent moving bed. The rotating cylindrical annulus and the countercurrent moving bed have been recently reviewed (Carr, 1993). Since then there have been no new developments on these two reactors, so only a short summary will be given here. There has been considerable progress on simulated countercurrent moving beds in the past few years, however, and these will be treated in more detail in this chapter. In addition, the development of pressure swing adsorptive reactors will be reviewed.

5.2.1 Rotating Cylindrical Annulus Chromatographic Reactors

The rotating cylindrical annulus chromatographic reactor consists of two concentric tubes, one of somewhat smaller diameter than the other, with an adsorbent and perhaps also a catalyst packed into the space between the tube walls. This assembly is rotated about its axis, and a feed stream enters one end of the bed through a stationary inlet. An inert carrier enters the same end of the bed, uniformly distributed about the annular area. This is illustrated in Figure 5.1.

A fluid element entering the bed with the feed stream is propelled in the axial direction by the carrier, and in the circumferential direction by the rotation. Reaction occurs, and the reactant(s) and product(s) describe helical paths through the bed, eluting at a particular angular position measured from the feed point. If the reaction components are selectively adsorbed, they elute from the annulus at positions determined by affinity for the solid, the least strongly adsorbed species at the smallest angular rotation, measured from the feed point, and others at larger angles in order of increasing adsorption affinity. Streams of separated or enriched components can be collected at the elution positions.

The homogeneous acid-catalyzed hydrolysis of aqueous methyl formate solutions was investigated at 298 K in a modification in which the annulus was stationary and the feed was rotated (Cho et al., 1980). The adsorbent

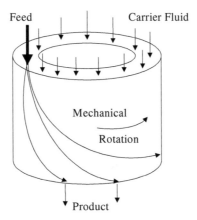

Figure 5.1:   The rotating cylindrical annulus chromatographic reactor.

was activated charcoal. Concentration profiles of the reaction products, methanol and formic acid, were determined by gas chromatography. No methyl formate could be detected in the reactor effluent, showing that 100% conversion was attained, whereas at the feed conditions the equilibrium conversion, the maximum attainable in a nonseparative reactor, would have been only 75%. Numerical simulations of reactor performance predicted 99% conversion, and complete separation of the two products. Complete separation was not experimentally observed because formic acid tailed badly on activated charcoal. The model predicted complete separation because the isotherm used did not account for tailing. Nevertheless, excellent performance of this rotating annulus reactor was experimentally observed, and predicted by the mathematical model.

The dehydrogenation of cyclohexane to benzene and hydrogen over a Pt/Al_2O_3 catalyst was also investigated in a rotating cylindrical annulus (Wardwell et al., 1982). For this gas–solid reaction system, studied in the vicinity of 473 K, the reactor was configured as a rotating annulus with a stationary inlet. Dispersion of the reactant and product bands was significant, preventing complete separation of the reaction components, and preventing 100% conversion. Nevertheless, the conversions were greater than the equilibrium conversion, again demonstrating superior performance of the adsorptive reactor compared with nonseparative reactors.

In the rotating annulus, the price paid for improved performance is inefficient utilization of the bed, as is evident from Figure 5.1, and the necessity for good rotating seals. In large-scale operations, the sliding friction from the seals would have to be overcome, and for gas–solid systems the seals would have to be precisely fabricated to minimize leaks. The stationary-bed/moving-feed configuration may be more desirable than the stationary-feed/moving-bed configuration, since in the former the seals would be on the drive shaft and would be smaller.

5.2.2   Countercurrent Moving-Bed Chromatographic Reactors

Another continuous flow configuration for carrying out simultaneous reaction and adsorptive separation is the countercurrent moving-bed chromatographic reactor (CMCR). The CMCR is a vertical tube through which the solid phase flows by gravity. The flow rate can be controlled by a constriction or by a valve at the bottom. The carrier fluid enters at the bottom and flows upward, counter to the solids. The feed is introduced at a position along the column. The solid and carrier flow rates can be adjusted so that the more weakly adsorbed chemical species move up the column with the carrier and can be removed near the top, while the more strongly adsorbed chemical species move down the column with the solid and can be removed near the bottom. This arrangement can be used as a continuous flow chromatograph for the separation of binary mixtures, or for separation of multi-component mixtures into two fractions. It can also serve as a chemical reactor. If a reversible reaction of the type $A = B$ occurs, A can be separated from B and high conversions of reactant can be attained. The CMCR is illustrated in Figure 5.2, where the reactant, A, is the more strongly adsorbed species.

The first mention of the CMCR in the literature appears to be due to Viswanathan and Aris (1974a), who published a mathematical model for a reactor with a first-order, irreversible surface reaction. Only the reactant was adsorbed. The model predicted that the reaction would go to completion in a finite length of bed (Viswanathan and Aris, 1974b). A series of papers by Takeuchi and Uraguchi (1976a,b, 1977a,b) further developed the concept of the CMCR. This was followed by theoretical investigations of the first-order reversible reaction, $A = B$, where adsorption of both A and B was described by Langmuir isotherms (Cho et al., 1982; Petroulas et al., 1985a,b). Axial concentration profiles for the dispersionless, adsorption equilibrium model exhibited shocks (concentration discontinuities), their number and position depending upon operating and boundary conditions. The model predicted conditions for which the reversible, equilibrium limited reaction could be driven to completion, and for which the product stream taken from the top of the CMCR would be only slightly contaminated with reactant. The model predictions have been verified by an experimental investigation of the hydrogenation of 1,3,5-trimethylbenzene, catalyzed by Pt/Al_2O_3, with Al_2O_3 as the adsorbent (Fish and Carr, 1989). The reaction was carried out in excess

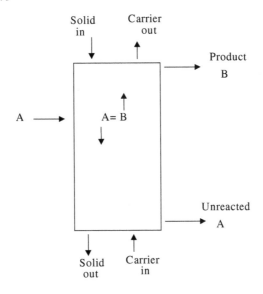

Figure 5.2:   The countercurrent moving-bed chromatographic reactor.

H_2 so that it would mimic $A = B$. Reaction conditions were found where the hydrogenated product, 1,3,5-trimethylcyclohexane, was recovered with only about 1% trimethylbenzene contamination, and the conversion was 90%, compared with the equilibrium conversion of 50% that would be expected in the absence of separation, all other conditions the same.

The solids flow in the CMCR presents a number of difficulties. In addition to the necessity to return solids to the top of the reactor, the descending solids are attrited by abrasion against one another. The resulting fines must be removed, and the decreasing particle size may change reaction rates. Furthermore, scale-up would require either large diameter columns, or multiple columns. The first approach would introduce a solids flow management scheme in order to avoid flow channeling, which would spoil the separation. The second approach would introduce additional complexity to ensure identical performance of each column in the bundle.

5.2.3   Simulated Countercurrent Moving-Bed Chromatographic Reactors

The solids-handling disadvantages of a CMCR can be avoided, and the advantages of countercurrency can be preserved, by employing simulated countercurrency (Ray et al., 1990). An arrangement for this is illustrated in

Figure 5.3. In a simulated countercurrent moving-bed chromatographic reactor (SCMCR) the solid phase does not move, but is held stationary in a fixed bed. The carrier enters one end of the bed and flows through it. A number of ports, serving as inlets and outlets, are equally spaced along the length of the bed. The feed enters the inlet at one end, for a specified time interval, then is advanced to the next, and then to the next, and so forth until the end is reached. When the feed is returned to the first inlet, the process is repeated as many times as desired. The product is removed from an outlet at a specified distance ahead of, or behind, the feed, depending on whether it is more or less strongly adsorbed than the reactant. This is described in the next paragraph. The outlet position is advanced in concert with the feed position. In contrast with the countercurrent moving-bed, where the solids flow past a stationary feed port, in the simulated countercurrent moving-bed the feed position moves past the bed. The inlet and outlet port movement is in the direction of carrier flow, and the frequency of feed port switching is selected so that the more strongly adsorbed species lag behind the feed port as they travel along the bed. The more weakly adsorbed species advance ahead of it at a flow rate somewhat slower than the carrier, which is chosen for its weak affinity for the adsorbent. For a reaction, $A = B$, the product can be removed at one of the outlets, and any unreacted A from the other. The reactive separation acts to break chemical equilibrium and leads to yield enhancement.

Steady-state material balances show how the carrier flow rate and the switching speed should be selected in order to effect the separation. The speed with which a single adsorbate obeying a Langmuir isotherm travels along the column in dispersionless flow is given by

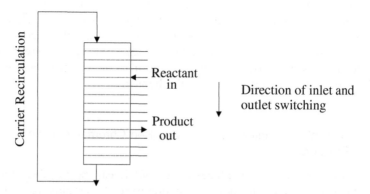

Figure 5.3:   The simulated countercurrent moving-bed chromatographic reactor.

$$V_f = \frac{U_g(1 - \sigma)}{1 + \dfrac{(1 - \varepsilon)NK}{\varepsilon(1 + CK)}} \qquad (5.1)$$

where ε is the interparticle void fraction, N is the total concentration of surface sites, K is the adsorption equilibrium constant, C is the fluid-phase concentration of the adsorbate, and σ is a parameter that governs the direction in which it travels with respect to the motion of the feed point. It is given by

$$\sigma = \frac{U_g(1 - \varepsilon)NK}{U_s\varepsilon(1 + CK)} \qquad (5.2)$$

where U_g is the speed of the carrier, U_s is the pseudo-solids speed. $U_s = x/t_s$, where x is the distance between adjacent inlets, and t_s is the feed switching time interval. If $\sigma < 1$, V_f is positive and the adsorbate is swept ahead of the feed. If $\alpha > 1$, V_f is negative and the adsorbate lags behind the feed. For a given adsorbent–adsorbate pair and given particle size, U_g and U_s may be selected so that $\sigma < 1$ for one and $\sigma > 1$ for the other. However, adsorbate concentration also plays a role. The decrease of σ as C increases will cause an adsorbate moving behind the feed ($\sigma > 1$) at a particular C to turn about and move ahead of the feed if C becomes large enough. At high enough concentrations, this component may overtake the other and spoil the separation.

The countercurrent moving bed operates in a steady-flow steady state, with axial concentration profiles that are time invariant. On the other hand, in the simulated countercurrent moving bed the axial concentration profiles are not time invariant. At each advance of feed position, the feed stream enters a region of the bed where existing concentrations were established while the feed was in the previous position. Entrance of fresh feed causes a disturbance and generates a transient which starts to relax to a steady state, but may or may not get there before the next feed advancement. The compositions of the effluent streams are time dependent, reflecting this transient behavior. Since the feed flow and composition are time invariant, and since the time interval between successive feed advancements is constant, the concentration transients are all identical. They are undamped periodic functions, and the reactor can be described as operating in a periodic steady state. This is a fundamental difference between the true countercurrent moving bed and the simulated countercurrent moving bed.

A mathematical model for a reversible, equilibrium-limited reaction, $A = B$, carried out in the SCMCR represented in Figure 5.3 predicts the high conversion and high product purity expected on the basis of the qualitative arguments stated above (Ray et al., 1994). The reactor is divided into

20 adsorption equilibrium stages, each stage having an inlet and an outlet port. Adsorption of A and B is described by a Langmuir isotherm, and a first-order reversible reaction occurs on the surface of a solid catalyst. The reactant entering the feed stage is uniformly distributed and is assumed to instantaneously come to adsorption equilibrium. The basis for choosing 20 stages is that the model simulates axial dispersion numerically by the width of the stages. The model equations describe transient concentrations both in the fluid phase and on the surface, adsorption, and chemical reaction. They are numerically integrated over the time interval between feed advancements by the fourth-order Runge–Kutta method, the concentration existing at the end of a feed interval serving as the initial condition for integration when the feed point is advanced. The feed and product removal locations are advanced when $t/t_s = 1, 2, 3, \ldots$, where t_s is the advancement time interval, or switching time. Since the reacting components are highly diluted by the carrier fluid, temperature changes caused by the enthalpy of reaction are neglected, and the energy balance is not in the model.

This model was used for numerical simulations of the SCMCR performance with 473 K data (reaction rate coefficients and adsorption isotherms) for the Pt/Al_2O_3 catalyzed hydrogenation of 1,3,5-trimethylbenzene (mesitylene, MES) to form 1,3,5-trimethylcyclohexane (TMC) at 1 atm total pressure with less than 1% MES in a 40% H_2-in-N_2 mixture. The data are from Fish and Carr (1989). At these conditions the equilibrium conversion is 40% (Egan and Buss, 1959).

Calculations of concentration profiles are shown in Figures 5.4 and 5.5 for an SCMCR 400 cm long with 20 stages and $t_s = 5$ s. The product stream

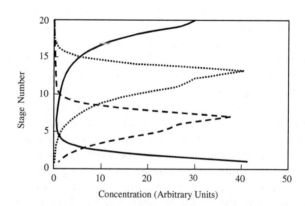

Figure 5.4: Numerical simulation of reactant concentration profiles in an SCMCR at τ ($\tau = t/t_s$) equal to 6 (-- -- --), 12 (.........), and 60 (———).

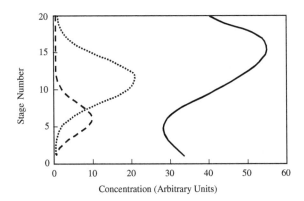

Figure 5.5: Numerical simulation of product concentration profiles in an SCMCR at $\tau(\tau = t/t_s)$ equal to 6 (– – – –), 12 (.........), and 60 (———).

consists of 15% of the vapor phase residing on the ninth stage downstream from the feed stage, where a 15% make-up flow enters. The profiles are those at the end of a feeding time interval, just prior to the advancement of the feed to the next inlet port. Functionally, the reactor has no top or bottom, and the inlet and outlet positions can be continually shifted around the reactor. Figure 5.4 shows the MES concentration profiles after 6, 12, and 60 switches of feed position. The feed location for these profiles is at stages 7, 13, and 1, respectively, since position 1 is the feed point at $t = 0$, and 60 switches corresponds to three complete cycles around the reactor.

In Figure 5.5 are the TMC profiles after 6, 12, and 60 feed switches, showing the growth of this reaction product over the start-up transient before the periodic steady state is reached. The MES profiles do not show a corresponding decrease because MES lost by reaction is replaced by the feed.

Comparison of Figures 5.4 and 5.5 shows that if TMC is removed 5–10 stages ahead of the feed, a highly enriched product will be obtained. The ninth stage gives the most enrichment, with the TMC contaminated by only 2.5% of MES. The conversion is 0.98. Numerical simulations showed that the conversion and purity depend upon the values of α for both TMC and MES, the switching time, the pseudo-solids speed, reactor length, fraction of flow withdrawn as feed, and reactant concentration in the feed. Conversions greater than 0.995 and purities greater than 0.99 were found for a range of operating conditions. This exceeds by far the equilibrium value of 0.4, the maximum attainable under nonseparative conditions, showing that reaction with adsorptive separation acts to break local chemical equilibrium, giving enhanced conversions and a high-purity product.

An alternative to the packed-tower SCMCR shown in Figure 5.3 is the multiple-column configuration of Figure 5.6, which is a more convenient arrangement for laboratory experimentation. Here, packed columns are connected in series for flow conduction between adjacent columns. The connections are also provided with valves for introduction of an external flow or for withdrawing a stream. Each column may be thought of as representing a segment of the Figure 5.3 arrangement between adjacent inlets and outlets. The number of columns is then a readily implemented variable, as columns can readily be inserted or removed as the application demands. It is clear that the multiple-column SCMCR is strictly analogous to the packed-tower SCMCR of Figure 5.3, and should offer the same performance.

The multiple-column SCMCR also attains a periodic steady state. The introduction of feed can be idealized as a step input of reactant into a regenerated feed column that is devoid of adsorbates. When the feed point is advanced the reactant concentration entering the former feed column is abruptly terminated and the reactant concentration takes a step down. The shape of the concentration front that sweeps through is governed by rates of adsorption, mass transfer from the fluid phase to the solid, and by dispersion in the fluid phase. There is an extensive literature on adsorber dynamics (Ruthven, 1984; Yang, 1987). This provides a framework for understanding the transients that lead to the SCMCR periodic steady state.

In laboratory-scale studies the adsorbent particle diameter is usually small, less than 0.1 μm, and external mass transfer is negligible. Also, unless the columns are unusually long or the flow rate unusually slow, dispersion will only play a minor role, and adsorption will be the dominant factor. The flow rate in an adsorber is usually selected so that adsorption rates are fast,

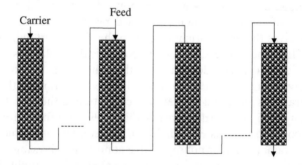

Figure 5.6:   Multiple-packed-column configuration of the SCMCR.

and the width of the mass transfer zone is narrow. Under these conditions adsorption can be described by an isotherm.

Isotherms can be classified as linear, favorable, and unfavorable (Yang, 1987). For a linear isotherm the surface concentration of the adsorbate is proportional to the fluid-phase concentration, and the isotherm does not influence the shape of the concentration front. In this case, dispersion and mass transfer will broaden the leading and trailing edge of the injected square wave somewhat. Favorable isotherms describe saturation of the surface, and a plot of surface concentration vs. fluid concentration is convex upward. The leading edge of a concentration wave for a component with a favorable isotherm is self-sharpening, and the trailing edge is self-broadening. This gives well-defined breakthrough times and broad desorption "tails." The most common favorable isotherm is the Langmuir isotherm. Unfavorable isotherms have concave plots of surface concentration vs. fluid concentration. Their influence on concentration waveforms is to broaden the front and sharpen the tail. Numerical simulation of a multicolumn SCMCR requires a numerical method that will faithfully reproduce the concentration waveforms, some of which are very steep. Fortunately, several approaches to this problem have been identified (Finlayson, 1992).

5.2.3.1 High-Temperature Reactions

Many chemical reactions of interest must be carried out at high temperature to attain adequate reaction rates, but adsorbents may decompose at high temperature, or may not provide enough adsorption selectivity for the required separation. In these cases, integrated adsorption and reaction in mixtures of catalyst and adsorbent, or homogeneous reactions in the presence of adsorbent, will not be possible. However, it is still possible to configure an SCMCR so that reaction and separation can be carried out. In this configuration each individual column of the SCMCR is divided into two segments: a fixed-bed reactor at the reaction temperature, followed by an adsorber at a lower temperature suitable for the separation.

The SCMCR consists of three or more of these pairs of reactors–adsorbers. It is important to note that each pair serves the same function as the integrated reactor–separator columns of the conventional SCMCR, and that the high-temperature reactor with its segregated reaction and separation is functionally identical to it in an overall sense. However, reaction with separation is no longer *locally* integrated, so there will be differences in performance. Figure 5.7 shows this concept. A difference between the reactors of Figures 5.6 and 5.7 is that when equilibrium-limited reactions are carried out in the former, local equilibrium is broken, and high per pass

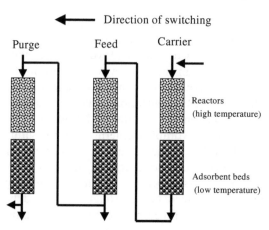

Figure 5.7: An SCMCR configured for high-temperature reactions. Each section of the SCMCR consists of a high-temperature reactor followed by a lower-temperature adsorbent bed.

conversions are possible. In the latter, the per pass conversion is equilibrium limited, but the reactor can be operated so that only the reaction product is withdrawn. If no reactant is permitted to escape, complete conversion is possible.

A modification of the high-temperature configuration consisting of only one reactor and three adsorbers is shown in Figure 5.8. In this arrangement, the feed continuously enters the reactor, but the reactor effluent is switched from one adsorber to the next in the direction of fluid flow, and a simulated countercurrent moving-bed separation is achieved. This arrangement appears to be a reactor followed by a separator, but closer consideration shows that it is an SCMCR that is functionally identical to the reactor in Figure 5.7, since reaction and separation occur in the feed segment in both arrangements. The advantage of the Figure 5.8 arrangement is the single reactor, which eliminates the necessity for running several reactors, all but one of which may be redundant at any particular time.

5.2.4 Pressure-Swing Adsorption Reactors

Pressure-swing adsorption is widely used in the chemical industry for carrying out separations by contacting gaseous process streams with adsorbents at high pressures, and then depressurizing to selectively desorb certain components. If conditions are chosen so that a chemical reaction

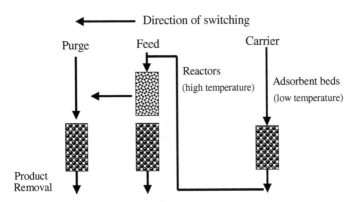

Figure 5.8: A high-temperature SCMCR modified to have only one chemical reactor.

occurs, then an integrated reaction–separation process results, and the reactor–separator unit is called a pressure-swing reactor (PSR). This concept was introduced by Vaporciyan and Kadlec (1987). They developed an isothermal, one-dimensional, dispersionless reaction with adsorption equilibrium model to describe pressure cycling of a single bed packed with a mixture of catalyst and adsorbent. A three-step cycle, consisting of high-pressure vapor feed, depressurization, and low-pressure exhaust, was investigated. Three solid catalyzed reactions, $A \leftrightarrow 2C$, $2A \leftrightarrow C$, and $A \leftrightarrow B/2 + C$, where all species were adsorbed via linear isotherms, and local chemical equilibrium was established, were treated. While the model predicted separation, conversions beyond chemical equilibrium were not observed because of the local equilibrium constraint. In subsequent work this constraint was removed (Vaporciyan and Kadlec, 1989), and it was found that for reversible reactions, conversions beyond equilibrium were predicted. An experimental study of the oxidation of CO to CO_2, an irreversible reaction, was also reported. Reaction rates were found to be higher in the pressure-swing reactor than in a conventional tubular reactor run at the same conditions. This was attributed to inhibition of the reaction rate by CO in the nonseparative reactor, and its absence, or at least lessened importance in the pressure-swing reactor due to removal of CO in the exhaust.

These pioneering studies demonstrated that the pressure-swing reactor is a viable adsorptive chemical reactor capable of providing better performance than nonseparative reactors. In subsequent research a number of configurations of the pressure-swing reactor have been investigated and several reactions have been examined. A number of patents have also been granted. Some of this work has been reviewed recently (Chatsiriwech et al., 1994). Lu and Rodrigues have performed a detailed modeling study of PSR for reactions of the type $A \leftrightarrow B + C$, such as dehydrogenation of ethane to ethylene (Lu and Rodrigues, 1994). They studied three key issues: effect of process parameters on PSR performance, evolution of process steady state from different initial conditions, and bed dynamics at cyclic steady state. More recently, Carvill and co-workers have presented an experimental demonstration of a five-step PSR concept for the reverse water–gas shift reaction (Carvill et al., 1996). For the same equilibrium conversion, they were able to reduce the operating temperature from 565 to 250°C, or an improvement in conversion from 9 to 36%. They stressed the importance of a countercurrent pressurization with product for optimum performance. UOP has patented two applications in this area: synthesis of methanol and isomerization of C_6 paraffins (Dandekar et al., 1996; Dandekar and Funk, 1998).

5.2.5   The Trickle-Bed Reactor

In contrast with the two-phase reactors described above, a three-phase adsorptive chemical reactor has been introduced (Westerterp and Kuczynski, 1987; Kuczynski et al., 1987). This is a gas–solid–solid trickle-flow reactor configured as a vertical tower with sections packed with catalyst pellets alternating with sections for heat removal. A powder, which is the adsorbent, slowly flows downward through the bed, and a chemically reacting gas flows upward, countercurrent to the powder. If the reaction product(s) is (are) selectively adsorbed by the powder, it (they) can be removed from the reactor, and if the reaction is reversible any equilibrium limitation can be overcome.

A one-dimensional, steady-state, mathematical model of this reactor has been developed and applied to methanol synthesis from CO and H_2 (Westerterp and Kuczynski, 1987). The model predicts complete conversion in a single pass for appropriate feed conditions and residence time. This may be contrasted with commercial reactors for methanol from syn gas, an equilibrium-limited reaction requiring separation and recycle of the unconsumed reactants even when carried out at high pressures in order to favor methanol as much as possible. With the model results for guidance, a reactor was built, and an experimental investigation was done (Kuczynski et al., 1987) using an amorphous silica alumina powder that selectively adsorbs methanol at reaction temperatures, in the vicinity of 500 K (Kuczynski et al., 1986). Complete conversion of reactants, introduced in the stoichiometric ratio, was observed at pressures from 5.0 to 6.3 MPa. The data were used for a process design, and an economic evaluation that appears favorable (Westerterp, 1988).

5.3 Issues in Adsorbent/Catalyst and Reactor Design

A number of issues need to be addressed when developing new adsorption with reaction processes. After screening several catalysts and adsorbents, detailed measurements are made of the physical characteristics of the system. These measurements include characterization of adsorption, equilibrium, mass transfer, and reaction kinetics. This section discusses the critical issues related to adsorption, reaction, and particle design.

5.3.1 Adsorption Issues

Pulse tests are a useful tool for screening potential adsorbents. However, for actual process modeling and scale-up, adsorption measurements must be made at solute concentrations close to those expected in the actual process. Furthermore, not just single component adsorption isotherms, but also multicomponent isotherms should be measured. The adsorbent chosen should not adsorb the preferred component too strongly. If the preferred component adsorbs strongly, desorbing it requires high desorbent usage, resulting in a loss of productivity. High desorbent flow rates result in short residence times, posing a potential imbalance for the reaction term, which may lead to incomplete conversion. Whenever possible, a dual-function solid should be used for both adsorption and catalytic functions. Such a solid allows maximization of adsorption capacity, which is critical for achieving good separation. Mass transfer to and from the adsorption site is also critical. In general the faster the mass transfer the better the process performance.

Identifying the controlling mass transfer resistance in the adsorbent, whether in the film, macropores or micropores, is important (Ruthven, 1984). Appropriate steps should be taken to reduce mass transfer resistance by controlling linear velocity, or changing particle size, crystallite size, or both. The sorbate–sorbate interactions inside microporous materials such as molecular sieves must be taken into account. This interaction can significantly alter mass transfer behavior at varying concentrations. Finally, the rate of mass transfer of intermediates, feed, and products must be considered. Their mass transfer rates have to match the requirements imposed on them by the reaction kinetics. For example, the rate of diffusion of *n*-hexane in H-ZSM-5 is eight orders of magnitude higher than 2,2-dimethylbutane and two orders of magnitude higher than 3-methylpentane (Karger and Ruthven, 1992).

5.3.2 Reaction Issues

A high-activity catalyst maximizes productivity because it occupies very little volume and so maximum separation capacity is available for enhancing conversion. Ideally, a single particle should contain both catalyst and adsorbent. Preferably, the desorbent should be one of the reactants. As reported by Mazzotti and co-workers in 1996, both of these conditions were satisfied. Using Amberlyst-15 as a catalyst and adsorbent and ethanol as a reactive desorbent, significant improvement in performance was achieved for esterification.

The effect of reaction intermediates also needs to be understood. For example, certain isomerization and cracking reactions form olefin intermediates. The olefins may adsorb on the adsorbent and take up adsorption capacity. In some cases, metals or sulfur on the catalyst may leave or migrate off the catalyst. This migration may adversely affect the adsorbent performance.

5.3.3 Particle Design

Whether the catalyst and adsorbent are a single particle or are on separate particles has important ramifications on the performance of the process. For the reaction of the type $A \leftrightarrow B + C$, where B is preferentially adsorbed over C, for maximum purity of C, the catalyst and adsorbent should be on a single particle. In this case, the product B, when produced at a catalyst site inside the pill is adsorbed in neighboring adsorbent sites, thus reducing the likelihood of B being present in the bulk solution. However, if the adsorbent and catalyst sites are on separate particles a reverse reaction is less likely to occur. Thus, the conversion in this case is maximized at the expense of purity. These scenarios are schematically shown in Figure 5.9. When a single product is formed, the adsorbent site should be removed from the catalyst site, and the reaction should be conducted in vapor phase so that the mass transfer is sufficiently fast to prevent back reaction of the product to the reactants. When a catalyst and adsorbent are on separate particles, understanding the packing characteristics of the two pills is necessary to avoid packing inhomogeneities and excessive voidage.

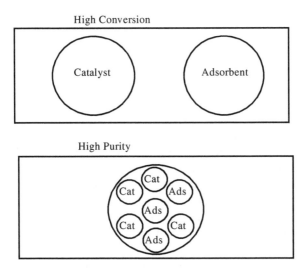

Figure 5.9: Particle design.

5.4 Applications

The SCMCR can be applied to good advantage to equilibrium-limited reactions, to certain reactions for which the conversion per pass is inherently small, and to reactions where maximizing the yield of an intermediate product is desired. In all of these cases it is the integration of separation with reaction that provides the benefit. The reactor can be run under conditions of optimum selectivity for the desired product while achieving high conversion. Selectivity is governed by chemistry, so the operating conditions can be chosen to give the chemical environment that maximizes selectivity. The reactor, on the other hand, ensures high conversion, since if only the product(s) is (are) removed, the reactant(s) must remain in the reactor until completely converted. In practice, it may not be possible to prevent some reactant losses. Several reactions that have been reported in the literature, and that illustrate these points, are discussed below.

5.4.1 Equilibrium-Limited Reactions

5.4.1.1 The Hydrogenation of 1,3,5-Trimethylbenzene

The production of 1,3,5-trimethylcyclohexane (TMC) by catalytic hydrogenation of 1,3,5-trimethylbenzene (mesitylene, MES) on Pt/Al_2O_3 has been investigated in a five-column SCMCR at 463 K and 473 K (Ray and Carr, 1995a). This gas–solid reaction was carried out in the presence of a large excess H_2 so that it has the stoichiometry of an $A = B$ reversible reaction. The adsorbent was 60/80 mesh Chromosorb 106, a porous polymer material, which was mixed with 10% by weight of the catalyst. A laboratory microcomputer was used to control flow switching through several two- and three-way solenoid valves, and to automatically sample the two reactor effluents for analysis by gas chromatography.

The five-column reactor is shown in Figure 5.10. The $MES/H_2/N_2$ feed enters column 2, where reaction occurs. Since the breakthrough time for TMC is about one-fourth of that for MES, TMC elutes first and a pure TMC-in-N_2 mixture is removed from port B until MES breaks through. Just before MES breakthrough, the feed is advanced to column 3 and port B and port A are simultaneously advanced to the next column. Figure 5.10 shows columns 1 and 2 serving as reaction columns, and columns 3, 4, and 5 being purged of remaining traces of adsorbed MES. This can be done at elevated carrier flow rates, if necessary. The objective is to remove MES from column 3 before it becomes the feed column, otherwise the TMC product taken from it will be contaminated with MES.

Figure 5.11 shows the time-dependent concentrations of TMC and MES at ports A and B for a 473 K experiment with a 5 min switching time. The open circles represent TMC and the filled circles represent MES. The hydrocarbons are highly diluted with N_2 carrier. The high dilution was deliberate since adsorption of both TMC and MES follows a Langmuir isotherm,

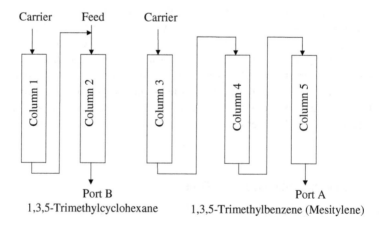

Figure 5.10: Five-column SCMCR for catalytic hydrogenation of mesitylene (MES).

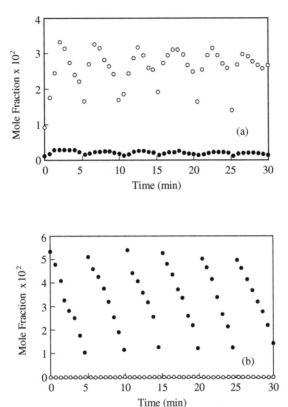

Figure 5.11: Concentration profiles (a) at port A of Fig. 5.10, (b) at port B of Fig. 5.10. Open circles, trimethylcyclohexane; filled circles, mesitylene.

which can be approximated by a linear isotherm at low concentration, and the reactor can be modeled by linear mathematics. The data were taken by sampling the effluent from each port at 40 s intervals. At port B, TMC breaks through quickly, goes through a maximum, and then decays to less than half of its peak value. There is also MES present. The MES eluting from port A is caused by desorption from the solid. It is the broadened "tail" of the waveform of MES exiting two columns behind the feed column. The existence of MES at port B shows that in this experiment MES purging at port A is incomplete. Note that there is very little TMC at port A. The shape of the TMC and MES concentration–time plot in Figure 5.11 clearly shows the periodic steady state that is characteristic of the SCMCR. Integration of the area under the concentration waves gave 81.6% conversion and at port B a TMC product purity of 92.9%. The best performance observed was 83% conversion and 96% purity. This is considerably better than the 40% maximum conversion that would be obtained in, for example, a fixed-bed catalytic reactor with the reaction taken to equilibrium, and demonstrates the capability of the SCMCR for giving enhanced performance for equilibrium-limited reactions.

A mathematical model of this reactor predicts 97% conversion and 98% TMC purity (Ray and Carr, 1995b). The difference between the experiments and the model calculations is due to the difficulty of completely desorbing MES. This is reflected in the experiments by the MES lost through port A, which manifests itself as lower conversion, and the MES eluting from port B, which decreases the product purity and also contributes to the loss of conversion. The Al_2O_3 has a distribution of binding energies with the aromatic ring. Desorption from high energy sites is slow, causing the "tail" of the MES concentration wave to be more extended than if the binding energy were monoenergetic. This is the source of the MES desorption tails seen in Figure 5.11, and explains why MES has not been completely purged from the feed column. The mathematical model, on the other hand, uses an adsorption isotherm with identical surface sites, and cannot account for the additional MES tailing. Better performance could presumably be obtained by placing another column in the reaction section, so that the tail seen at port A in the five-column reactor would be in the added sixth column and less MES would be lost. Alternatively, increasing the carrier flow in the purge section would sweep more MES forward into the reaction section, also reducing losses.

5.4.1.2 Methanol Synthesis

A mathematical model for catalytic production of methanol from synthesis gas has been developed and applied to adiabatic and isothermal versions of the SCMCR (Kruglov, 1994). In these multicolumn SCMCRs, H_2 was in 100% stoichiometric excess over CO, and served as the carrier gas. The model equations consisted of a fixed-bed transient species balance for each of the three components, and a transient energy balance equation. Radial uniformity of concentration and temperature was assumed. The species balances accounted for convection, axial dispersion, and the chemical reaction rate, the last by a pseudohomogeneous rate expression. The energy balance accounted for gas-phase heat convection, heat generation by reaction, and heat generation by adsorption. Methanol was the only component that adsorbed on the solid. The numerical solution of these equations employed an implicit central finite-difference method for spatial derivatives. To cope with the steep concentration profiles that propagate through the columns, adaptive grid generation, in which the space grid was recalculated at each time step, was used. Calculations were done for a reactor inlet temperature of 500 K and an inlet pressure of 60 bar (6 MPa). Plots showing the axial evolution of concentration, temperature, and flow rate during a switching interval were presented.

Two configurations of the adiabatic reactor were studied. One was a four-column arrangement with three reaction columns and one purge column. The CO and H_2 feed entered the first of the three reaction columns, and the purge was the first column upstream of the feed. CO and H_2 are assumed not to adsorb, and travel ahead of the feed advancement, while adsorbed methanol lags it and is recovered from the purge column. The CO conversion was only 0.36, since the CO and H_2 residing in the feed column are lost, eluting with methanol when the feed point is advanced and the former feed column becomes the purge column. This CO loss was avoided by adding one more column between the feed and the purge. In this five-column arrangement, the reaction section has been increased from three to four columns, and the CO and H_2 in the feed column are left one reactor section behind upon a feed switch. It is only necessary to select the switching time so that this CO and H_2 are swept out before this column becomes the purge at the next feed switch. With this arrangement, 95% conversion was obtained.

In the isothermal SCMCR (Kruglov, 1994), the adsorbent and catalyst are not mixed. Five fixed-bed adsorbers were used in a simulated counter-current mode, and three of these were connected in series with three adjacent adsorbers. The reactors were stationary with the feed always entering the first one. In this reactor, simulated countercurrency was achieved by moving each adsorber one position in the direction of fluid-phase flow at the end of each switching interval, thus placing a different adsorber in series with each of the three reactors. Methanol was purged from the adsorber two columns upstream from the first reactor, and the adsorber between this one and the first reactor was used to keep CO and H_2 in the SCMCR, as for the five-column adiabatic reactor. Isothermal, rather than adiabatic, operation was necessary, since the enthalpy of reaction could not be taken up by recirculating solids, but required external cooling of the fixed-bed reactors. It was found that CO conversions could be increased to 97–99% in the isothermal reactor.

5.4.1.3 The Esterification of Acetic Acid

The formation of ethyl acetate by acid-catalyzed esterification of acetic acid and ethanol has been investigated in a multicolumn liquid–solid-phase SCMCR (Mazzotti et al., 1996). The esterification was catalyzed by Amberlyst-15, a polystyrene–divinylbenzene resin containing sulfonic acid groups which also served to selectively adsorb the four reactive components, water, ethanol, acetic acid, and ethyl acetate. An approach to reactor design was described in which adsorption equilibrium data and kinetic data were used as a basis. It was shown that these were sufficient to characterize the SCMCR. Experimental adsorption data were obtained by contacting the resin with either single chemical species or a binary mixture. Partitioning of the reactive components between the liquid and solid phases was not described by adsorption isotherms, but by considering the swollen resin to be a homogeneous gel phase in contact with the liquid. The chemical activities of species in the gel were calculated by the extended Flory–Huggins theory, and in the liquid by the group contribution method. The gel and liquid activities of each component were then equated, making a phase-equilibrium model. Kinetic data were obtained from batch reactor experiments.

Experiments were then done with a chromatographic reactor, which was a single tubular fixed-bed reactor packed with Amberlyst-15 resin, and initially loaded with pure ethanol. When fed with acetic acid, there was a transient period during which reaction occurred and the weakly adsorbed product, ethyl acetate, eluted with ethanol, water and acetic acid being absent. At later times the more strongly adsorbed water and acetic acid also eluted from the reactor, eventually giving the chemical equilibrium composition. The performance during the transient period demonstrated the feasibility of simultaneous reaction and separation, and provided data necessary for designing the SCMCR. An eight-column SCMCR was then assembled. It consisted of a five-column reaction section, at the end of which a mixture of ethyl acetate and ethanol was removed, a two-column water stripping section, and a one-column acetic acid stripping section. Ethanol served as the carrier fluid, as well as being a reactant. Conditions were found in which 100% conversion of acetic acid, the limiting reagent, was obtained, and the two products were completely separated.

5.4.1.4   The Synthesis and Hydrolysis of Methyl Acetate

Methyl acetate is used in the manufacture of acetic anhydride, which is primarily used in making acetate films, such as cellulose acetates. The reaction is

$$CH_3OH + CH_3COOH \leftrightarrow CH_3COOCH_3 + H_2O \qquad (5.3)$$
$$\text{methanol} \quad \text{acetic acid} \qquad \text{methyl acetate} \quad \text{water}$$

Methyl Acetate Synthesis. Funk and co-workers demonstrated the synthesis of methyl acetate from methanol and acetic acid using a liquid-phase simulated moving-bed (SMB) adsorption with reaction (SMBAR) process (Funk et al., 1995). Figure 5.12 shows a schematic of this process in terms of an equivalent true moving bed process. The process consists of eight discrete beds, which were divided into three different zones. Amberlyst-15 was used as both an adsorbent and a catalyst. The process is described in terms of an equivalent true moving bed. The feed consists of pure acetic acid, and the second reactant, methanol, is used as a desorbent. Of the two products, water is more strongly adsorbed and methyl acetate less strongly adsorbed. A raffinate stream containing predominantly methyl acetate, and an extract stream containing mostly water and methanol are withdrawn. By continuously switching the ports of the feed and effluent streams in the direction of the fluid flow, a simulated movement of solid is created that is countercurrent to the direction of fluid flow.

Acetic acid is fed into the center of the reactor, where it encounters a downward-flowing liquid stream containing methanol in the presence of Amberlyst-15 to form methyl acetate and water. Amberlyst-15 preferentially adsorbs water and carries the water upwards. The separation of methyl acetate from water causes a complete conversion of acetic acid to methyl acetate. As the adsorbent travels upward past the feed point, it encounters a downward stream of methanol, which strips the water from it. Thus, an extract stream that contains mostly methanol and water is withdrawn. The remaining water from the Amberlyst-15 is stripped in the topmost zone, where it encounters more methanol. Thus, when the adsorbent "travels" around to the bottom of the reactor, it has little or no water in it. This minimizes any back reaction of methyl acetate and water, thus preventing any reduction in conversion. Thus, the raffinate contains mostly methyl acetate and methanol.

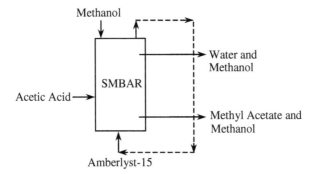

Figure 5.12: Methyl acetate synthesis by adsorption with reaction.

Fixed-bed experiments under conditions similar to the SMB adsorber reactor conditions revealed an equilibrium conversion of 90%. In this process, 100% conversion of acetic acid was achieved with a 99.8% recovery of methyl acetate. Unfortunately, methanol and methyl acetate form an azeotrope so that the downstream recovery of methyl acetate and recycle of methanol is expensive. The market for methyl acetate is limited, and the existing process for methyl acetate synthesis by Eastman Chemicals Company is quite competitive. Consequently, this process was not commercialized.

Methyl Acetate Hydrolysis. The hydrolysis of methyl acetate to methanol and acetic acid is an important recovery step in a proposed flow scheme to produce secondary linear alcohols (Funk et al., 1996a). Typically, an acid catalyst is necessary to perform this reaction. Amberlyst-36 resin is one such solid acid catalyst, which can be used up to 140°C. This reaction is carried out in the liquid phase. Amberlyst-36 preferably adsorbs methanol and rejects acetic acid. This reaction–separation can be seen in the pulse test shown in Figure 5.13. A feed pulse containing methyl acetate and water was injected to a stream of water and glyme (ethylene glycol dimethyl ether). Glyme is added as a solvent to improve miscibility between methyl acetate and water. The reaction products, methanol and acetic acid, elute as two distinct peaks with good separation between them. Thus, Amberlyst-36 showed promise as both a catalyst and an adsorbent. A bench-scale pilot plant run was performed using an eight-bed Sorbex™ plant. A schematic of the zone configuration is shown in Figure 5.14. A schematic of the equivalent true moving-bed process is shown in Figure 5.15. By continuously changing the positions of the inlet and outlet streams, countercurrent movement of the solid was simulated with respect to the liquid flow. No flow was

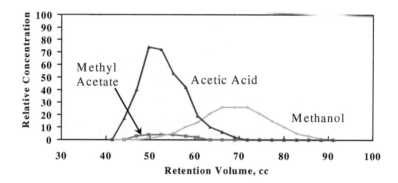

Figure 5.13:   Methyl acetate hydrolysis pulse test.

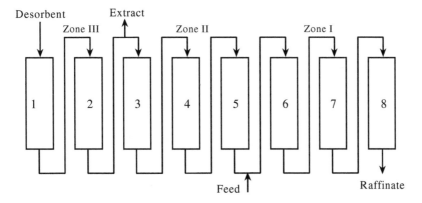

Figure 5.14: Zone configuration for methyl acetate hydrolysis.

allowed between Zone III and Zone II. The process is described in terms of an equivalent true moving bed. Solid containing Amberlyst-36 resin, which acts as both an adsorbent and a catalyst, moves from bottom to top countercurrent to the flow of liquid, which moves from the top to the bottom.

A typical concentration profile is shown in Figure 5.16. The feed is introduced at the bottom of Bed 4. The extract is removed at the top of Bed 6, and the raffinate at the bottom of Bed 1. The reaction takes place predominantly in Zone I (between feed and raffinate points). The liquid flows downward (from right to left in the figure) and the simulated movement of the solid is upward (from left to right). Methanol so produced is carried upward with the (simulated) upward-moving solid, along with some methyl

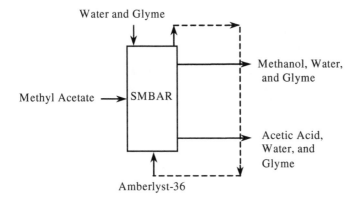

Figure 5.15: Methyl acetate hydrolysis by adsorption with reaction.

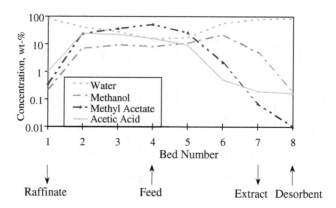

Figure 5.16: Experimental concentration profile.

acetate, although acetic acid is almost completely rejected. This mixture enters Zone II. Here a mixture of water and glyme desorbs methyl acetate off the resin.

Consequently, along Zone II (from left to right) in Figure 5.16, more and more methyl acetate is removed. The solid traveling beyond Zone II contains predominantly methanol. The methanol is stripped from the solid by a stream of water and glyme, which is introduced at the top of Zone III as a desorbent. An extract stream containing predominantly methanol and water is recovered.

5.4.1.5 *Process Studies*

The relative flow rates in various zones, the cycle time, and the feed rate strongly affect the performance of an SCMCR. These effects are demonstrated for the methyl acetate hydrolysis case:

- *Effect of Zone II rate.* As the Zone II flow rate increases, increasingly more methyl acetate is removed from the adsorbent. A further increase in flow rate leads to the desorption of methanol from the adsorbent. This methanol is pushed into Zone I. This results in loss of conversion because methanol reacts with acetic acid in Zone I to form methyl acetate. This methyl acetate is lost with the raffinate. Thus, an increase in the Zone II flow rate leads to an increase in purity but a decrease in conversion (Figure 5.17).
- *Effect of desorbent rate or Zone III rate.* If the desorbent flow is not sufficient, some methanol stays on the solid past Zone III, and

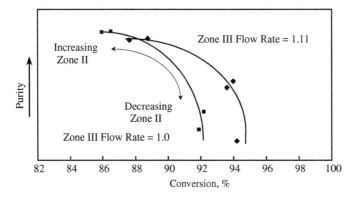

Figure 5.17: Effect of Zone II and Zone III flow rate.

"travels" around and enters the bottom of Zone I. This methanol reacts with the acetic acid in the raffinate to form methyl acetate, thus leading to poor conversion. An increase in desorbent rate causes the curve to shift from the bottom-left corner of the graph in Figure 5.17 to the top-right corner. However, the recovery of desorbent downstream increases utility costs.

• *Effect of feed rate.* Figure 5.18 shows purity–conversion curves for three different feed flow rates. The capacity of the adsorbent is clearly limited and a decrease in the value of the feed flow rate value from 2 to 1.5 and finally to 1 leads to a dramatic improvement of performance. Of course, a decrease in feed rate corresponds to a reduction in productivity of the unit.

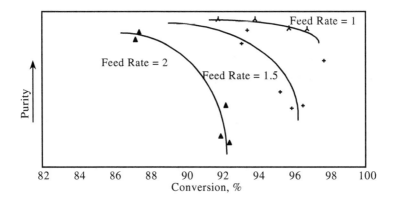

Figure 5.18: Effect of feed rate.

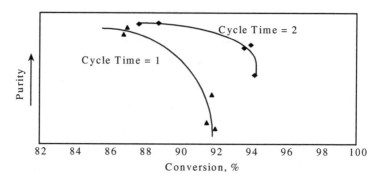

Figure 5.19: Effect of cycle time.

- *Effect of cycle time.* The process is limited by the overall rate of mass transfer of the species to and from the reactive adsorbent. Figure 5.19 illustrates this effect. When the cycle time is increased by a factor of two, the conversion at the same purity level increases by four points. However, a longer cycle time translates into lower productivity. The best conversion under these conditions is 91.5% at 58% purity and the best purity is 95% at a conversion of 87%. Fixed-bed experiments were performed to quantify equilibrium conversion under conditions similar to those at which a countercurrent simulated-moving-bed process was run. Experimentally measured fixed-bed equilibrium conversion for methyl acetate, under conditions similar to those for the countercurrent simulated-moving-bed process, was 69%. By using SCMCR process, the conversion was enhanced from 69% to 87%.

5.4.2   Selectivity-Limited Reactions

5.4.2.1   *The Isomerization of Light Paraffins*

New reformulated gasoline specifications have been, or will be, forcing refiners to reduce the amount of olefins and aromatics in gasoline. Iso-paraffins in the C_5 and C_6 range can be used as a substitute but straight-run naphtha contains relatively few branched isomers of C_6 paraffins. Isomerization of light paraffins is used to convert normal and mono-branched paraffins to higher-octane, dibranched paraffins. The traditional recycle isomerization process is depicted in Figure 5.20. The isomerization of *n*-hexane is usually carried out using an acid catalyst with a metal in the

Reaction

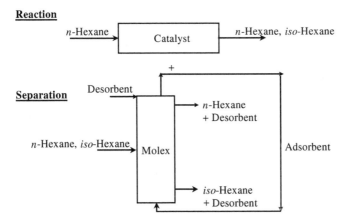

Figure 5.20:   Traditional recycle isomerization process with adsorption separation.

presence of hydrogen. The acid function is due to the presence of a zeolite or chloride on the catalyst.

The effluent is separated using fractionation or an adsorptive separation process such as Molex^{TM} using a zeolitic adsorbent. n-Hexane is extracted and recycled back to the process. The raffinate contains methylpentanes and dimethylbutanes, which are collected as a high-value product. Thus, this new process that could increase conversion beyond the equilibrium would be attractive (Funk et al., 1996b; Dandekar et al., 1998a).

The network of reactions can be represented as an equivalent reaction of the type $A \leftrightarrow B \leftrightarrow C$. The objective is to increase conversion of A and B to C. The catalyst used in this example is a zeolite with an alumina binder. To overcome the equilibrium, an adsorbent is needed which will selectively adsorb A and B, and reject C. After a number of screening tests, a suitable adsorbent was identified. Figure 5.21 shows a pulse test in which a feed pulse containing a mixture of C_6 isomers was fed in a steady stream of desorbent over a fixed bed containing a mixture of adsorbent and catalyst particles. The elution profiles clearly show a good separation between methylpentanes, n-hexanes, and dimethylbutanes. The process is run in vapor phase to facilitate mass transfer.

Figure 5.22 shows the schematic of the process. The simulated movement of solid is from the bottom to the top. The vapor flow is from top to bottom in a countercurrent direction to the flow of solid. As shown in the figure, the beds can be divided into a number of zones. Zone I is referred to as the reaction zone, Zone II and Zone III are the purification and recovery zones, respectively, and Zone IV (if present) is the buffer, or the "dead" zone.

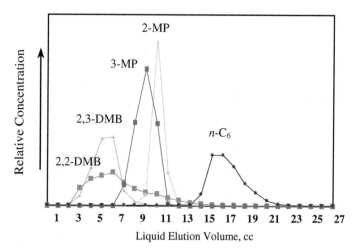

Figure 5.21: C₆ paraffin pulse test (2,2-DMB = 2,2-dimethylbutane; 2,3-DMB = 2,3-dimethylbutane; 2-MP = 2-methylpentane; 3-MP = 3-methylpentane).

The feed containing predominantly *n*-hexane is introduced at the top of Zone I. It flows with the downward vapor flow to encounter a mixture of catalyst and adsorbent moving in the opposite direction. The feed isomerizes to a mixture of *n*-hexane, methylpentanes, and dimethylbutanes.

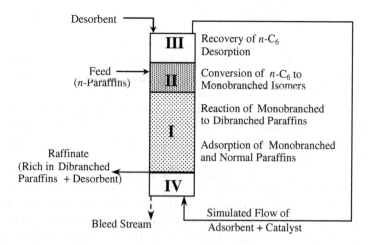

Figure 5.22: Schematic of countercurrent adsorption with reaction process for paraffin isomerization.

The methylpentanes and *n*-hexane are preferentially adsorbed and carried upward to Zone II. Little of the dimethylbutanes is adsorbed. The remaining dimethylbutanes are recovered as a high-octane raffinate product at the bottom of the reactor. In Zone II and Zone III, the adsorbent encounters a downward-flowing vapor stream containing desorbent. The desorbent initially strips the methylpentanes and later the *n*-hexane from the adsorbent. The desorbed *n*-hexane and methylpentanes come in contact with the catalyst to further react to dimethylbutane. Ultimately, a substantial fraction of the *n*-hexane is converted. An extract stream containing predominantly methylpentanes can be withdrawn just above Zone II.

Depending on the strength of the desorbent, some *n*-hexane can remain on the adsorbent and "travel" around and appear at the bottom of Zone I (Figure 5.22). This leads to a loss in conversion and decrease in purity. Thus stripping the *n*-hexane from the adsorbent before it leaves Zone III is important. One way that stripping can be accomplished is by increasing the desorbent flow. However, too high a flow reduces the residence time and does not allow the conversion of methylpentanes to dimethylbutanes in Zone I. Also, too high a desorbent flow rate will not allow adsorption of *n*-hexane in Zone I and will force *n*-hexane down into the raffinate. Thus, an optimum desorbent flow rate can be reached beyond which no improvement in conversion can be obtained. An alternative is to use a buffer zone between the top of Zone III and bottom of Zone I. If no bleed stream is taken at the bottom of Zone IV, then a dead zone exists between Zone I and Zone III. Here the unreacted *n*-hexane is desorbed (because of concentration gradients) and reacts to form an equilibrium mixture of C_6 isomers, which has a substantially higher octane value than pure hexane. Alternatively, a bleed stream may remove the *n*-hexane at the bottom of Zone IV and recycle it as feed.

A reaction with adsorption process for paraffin isomerization is fairly complex and requires understanding the effect of a number of parameters. A process this complex cannot be understood by performing experiments alone. A detailed modeling study was simultaneously undertaken to understand the effect of various process parameters. In particular, the effects of relative rates of mass transfer and reaction, the catalyst-to-adsorbent ratio, zone configuration, and the zone flow rates were studied.

Detailed mathematical models were developed both for the fixed-bed test to extract adsorbent parameters and the continuous countercurrent process. A dispersed plug flow model with finite mass transfer resistance was developed. Adsorption kinetics were defined by the Langmuir–Hinshelwood–Hougen–Watson (LHHW) rate expression. Figure 5.23 shows a typical concentration profile predicted by the model. *n*-Hexane, which is fed in at the center, is rapidly converted to methylpentanes. Any

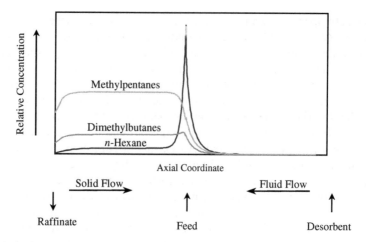

Figure 5.23: Numerical simulation of paraffin isomerization process.

unreacted *n*-hexane is refluxed back because the desorbent is flowing countercurrent to the movement of the solid. Because of the poor mass transfer characteristics of the adsorbent, the separation of dimethylbutanes from methylpentanes was slower than the rate of backward reaction and as a result the reaction of *B* to *C* could not exceed equilibrium. Thus, the profiles for methylpentane and dimethylbutane are parallel in the lower part of the bed.

A pilot plant was built, which was made up of a number of discrete fixed beds with a suitable system of valves to allow continuous switching of inlet and outlet ports to generate a countercurrent movement of solid with respect to the liquid. An experimental concentration profile is shown in Figure 5.24. Table 5.1 compares the best performance of the pilot plant with a vapor-phase fixed-bed operation under identical conditions of temperature and pressure. Also shown is the performance of a fixed-bed operation under commercial conditions.

To enhance mass transfer, the process was run at low pressure in the vapor phase. A 6% increase in conversion of 2,2-dimethylbutane was achieved compared to fixed-bed conversion. Because the reaction of methylpentanes to 2,3-dimethylbutane is very fast, no improvement in equilibrium was observed. Commercially, the process runs at high pressure, resulting in higher equilibrium conversion. The SMB process thus failed to deliver a performance better than best current commercial performance. Even though the process in its current state is a technical success, it still requires more optimization to make it a commercial success.

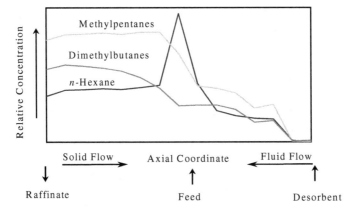

Figure 5.24: Experimental concentration profile for paraffin isomerization.

5.4.2.2 Methane Oxidative Coupling

Methane is the major component in natural gas, which is in abundance worldwide. There has been considerable interest recently in developing technology to convert methane and other natural gas constituents into higher-value chemicals (Hutchings and Scurrell, 1992; Lunsford, 1995). One of the routes that has received much attention is methane oxidative coupling, a reaction of methane catalyzed by metal oxides in the presence of oxygen, yielding ethane, ethylene, and carbon dioxide and water.

$$2CH_4 + \tfrac{1}{2}O_2 = C_2H_6 + H_2O$$
$$2CH_4 + O_2 = C_2H_4 + 2H_2O \qquad (5.4)$$
$$CH_4 + 2O_2 = CO_2 + 2H_2O$$

Table 5.1: Product Composition from SMB and Fixed-Bed Operation.

Product	Fixed Bed (wt%)	SMB (wt%)	Commercial Fixed Bed (wt%)
2,2-Dimethylbutane	22	28.5	32.1
2,3-Dimethylbutane	9.4	9.4	9.9
2-Methylpentane	30.7	28.3	31.7
3-Methylpentane	19.2	17.9	16.2
n-C_6	14.7	12.9	10.1
RON of Product	72	75	79

To obtain high selectivity for ethane and ethylene, the reaction must be carried out with excess methane, frequently as much as 20- to 50-fold excess to gain C_2 selectivity as great as 90%. Under these conditions, methane conversion is small and the C_2 yield is too low to be of commercial interest. The best C_2 yields reported are 20–25%.

When this intrinsically low-conversion-per-pass reaction has been carried out in an SCMCR, C_2 yields of about 50% have been obtained (Tonkovich et al., 1993; Tonkovich and Carr, 1994; Kruglov et al., 1996). The reaction occurs at high temperature, in the vicinity of 1000 K, so the high-temperature SCMCR design, with sections consisting of a reactor and a lower-temperature adsorber, must be used. One of the advantages here is that the C_2 products, which are more easily oxidized than methane, are immediately removed from the reactive environment when they enter the adsorber, minimizing their loss by the second step of a consecutive reaction. Another factor leading to good performance is that the SCMCR operates at high CH_4/O_2 and so there is excellent C_2 selectivity. Although the reactant concentration ratio in the SCMCR is favorable for high C_2 selectivity, the feed composition is stoichiometric, since it replaces only the reactants converted in a reactor–separator section per pass. The final factor leading to higher C_2 yields is the good overall methane conversion, up to 70–80%, that is obtained by repeatedly contacting the excess methane with the catalyst.

5.5   Process Evaluation

We will discuss the evaluation of a unit operation in which simultaneous adsorption with reaction takes place in the context of how it fits in the overall process flowsheet and what is its economic impact on the overall process economics.

5.5.1   Impact on Process Flowsheet

The following issues need to be taken into account to understand the impact of introducing a new reaction with adsorption unit operation in an existing process flowsheet:

- In simulated moving-bed adsorption processes, a desorbent is needed to recover the strongly adsorbed species and to regenerate the adsorbent. In cases where the desorbent is not one of the reactants, additional downstream separation may be necessary. This would require additional separation units downstream such as distillation columns, or high-pressure separators for when the desorbent is carrier gas.

- If a conventional fixed-bed reactor is replaced by a pressure-swing adsorption with reaction process, it may be necessary to add the capability to recover products from the purge medium. This would involve adding new pieces of equipment such as compressors, valves, holding tanks, etc.
- The issue of operability is really important. In a mature industry the operating businesses will be reluctant to switch to a multistep, unsteady-state process that has many more moving (switching) parts. The process should be designed to balance maximum productivity, complexity, and operability of the unit.

5.5.2 Economic Impact

A new unit operation that is based on reaction with adsorption can be used to replace an existing reactor. In most cases, one is trading in the simplicity of a sequential reaction with separation, with the complexity of single unit operation in which these functions are combined. Thus, the benefits that can be realized due to simultaneous reaction with separation have to outweigh additional capital and operating costs that the new unit operation may impose on the process. In particular:

- Is there a large enough differential in value between the feed and products in the process to justify the added complexity of the new unit operation?
- Hidden additional capital costs such as the increase in the size of reactors can arise when adsorbent is added. For example, in PSR processes, only one of the two or more beds is actually generating products. The other beds are undergoing recovery and regeneration. Is the additional capital cost of such a system compared to a single fixed-bed reactor justified by an increase in productivity?
- New operating costs might be imposed because of the use of a desorbent or purge. These might result in additional downstream separation steps to recover the desorbent. Does the increase in conversion justify the additional utility costs in the distillation columns used to recover the desorbent?
- The process has to be judged in comparison with the best-in-class current operating process. For optimum adsorption the temperature of the process may need to be lowered; and to maximize mass transfer rate the pressure might have to be lowered. Under these conditions, the conversion for the reactive adsorption process might exceed that for a fixed-bed operation. However, these may not be the optimum conditions for the fixed-bed processes. Under optimum fixed-bed con-

ditions, the equilibrium conversion might even exceed the conversion obtained using reactive adsorption.

5.6 Conclusions

Reactive chromatography has been known for the past 40 years. The advantages and applicability of continuous adsorption with reaction processes have been discussed. Continuous operation can be achieved through a number of configurations. These include trickle bed, true moving bed, annular chromatography, SCMCR, and PSR. As part of process development it is important to select the right adsorbent and catalyst for the system. The key issues in adsorbent selection and design have been discussed.

We have described a number of diverse applications of both equilibrium-limited and selectivity-limited reactions. Equilibrium-limited systems such as hydrogenation of 1,3,5- trimethylbenzene, hydrolysis and synthesis of esters, and methanol synthesis show improvement in conversion over equilibrium. We have also demonstrated improvement in selectivity for reactions such as isomerization of paraffins and methane coupling. They cover operations both in liquid- and vapor-phase operation.

However, we do not have a commercial process. For a process to be both technically and commercially successful, the keys for success are: (1) the reaction and adsorption conditions are closely matched; (2) the catalyst and adsorbent are one; and (3) one of the reactants can be used as purge. In this regard the synthesis of acetates has been the most successful story.

In general, adsorption with reaction technology does not have wide-ranging applications but will be useful in a high-value-added niche application. Successful commercialization would require simultaneous development of catalyst and adsorbent. Furthermore, the unit operation development should be undertaken only after the impact of the unit operation on the process flowsheet and its economic potential is examined.

Symbols

C	Fluid-phase concentration, kgmol m^{-3}
K	Adsorption equilibrium constant, m^3 kgmol^{-1}
N	Total concentration of surface sites, kgmol m^{-3}
t_s	Time step, s
U_g	Speed of carrier gas, m s^{-1}

U_s	Pseudo-speed of solid, m s^{-1}
V_f	Effective adsorbate velocity, m s^{-1}
x	Distance between adjacent inlets, m

Greek

| σ | Dimensionless parameter |
| ε | Intraparticle void fraction |

References

Aris, R., *Elementary Chemical Reactor Analysis,* Prentice-Hall, Englewood Cliffs, NJ, 1969, p. 72.

Basset, D.N. and H.W. Habgood, *J. Phys. Chem.,* **64**, 769, 1960.

Carr, R.W., in *Preparative and Production Scale Chromatography,* P. Ganetsos and P.E. Barker, Eds., Marcel Dekker, New York, 1993, p. 421.

Carvill, B.T., J.R. Hufton, M. Anand, and S. Sircar, *AIChE J.,* **42**, 2765, 1996.

Chatsiriwech, D., E. Alpay, L.S. Kirshenbaum, C.P. Hull, and N.F. Kirkby, *Catal. Today,* **20**, 351, 1994.

Cho, B.K., R.W. Carr, and R. Aris, *Sep. Sci. Technol.,* **15**, 679, 1980.

Cho, B.K., R. Aris, and R.W. Carr, *Proc. R. Soc. London,* **A383**, 147, 1982.

Dandekar, H.W. and G.A. Funk, U.S. Patent 5,811,630, 1998.

Dandekar, H.W., G.A. Funk, J.D. Swift, and R.T. Maurer, U.S. Patent 5,523,326, 1996.

Dandekar, H.W., G.A. Funk, S.H. Hobbs, M. Kojima, R.D. Gillespie, H. Zinnen, and C.P. McGonegal, U.S. Patent 5,744,683, 1998.

Egan, C.J. and W.C. Buss, *J. Phys. Chem.,* **63**, 1887, 1959.

Finlayson, B.A., *Numerical Methods for Problems with Moving Fronts,* Ravenna Park, Seattle, 1992.

Fish, B.B. and R.W. Carr, *Chem. Eng. Sci.,* **44**, 1773, 1989.

Funk, G.A., J. Lansbarkis, and A.K. Chandhok, U.S. Patent 5,405,992, 1995.

Funk, G.A., H.W. Dandckar, and S.H. Hobbs, U.S. Patent 5,502,248, 1996a.

Funk, G.A., H.W. Dandekar, M. Kojima, and S.H. Hobbs, U.S. Patent 5,530,172, 1996b.

Hutchings, G.J. and M.S. Scurrell, in *Methane Conversion by Oxidative Processes,* E.E. Wolf, Ed., Van Nostrand Reinhold, New York, 1992, p. 201.

Karger, J. and D.M. Ruthven, *Diffusion in Zeolites and Other Microscopic Solids,* Wiley & Sons, New York, 1992, p. 489.

Kruglov, A.V., *Chem. Eng. Sci.,* **49**, 4699, 1994.

Kruglov, A.V., M.C. Bjorklund, and R.W. Carr, *Chem. Eng. Sci.,* **51**, 2495, 1996.

Kuczynski, M., A. van Ooteghem, and K.R. Westerterp, *Colloid Polym. Sci.,* **264**, 362, 1986.

Kuczynski, M., M.H. Oyevaar, R.T. Pieters, and K.R. Westerterp, *Chem. Eng. Sci.,* **42**, 1887, 1987.

Lu, Z.P. and A.E. Rodrigues, *AIChE J.,* **40**, 1118, 1994.

Lunsford, J.H., *Angew. Chem., Int. Ed. Engl.,* **34,** 970, 1995.

Mazzotti, M., A.V. Kruglov, B. Neri, D. Gelosa, and M. Morbidelli, *Chem. Eng. Sci.,* **51,** 1827, 1996.

Petroulas, T., R. Aris, and R.W. Carr, *Comput. Maths. Applic.,* **11,** 5, 1985a.

Petroulas, T., R. Aris, and R.W. Carr, *Chem. Eng. Sci.,* **40,** 2233, 1985b.

Ray, A. and R.W. Carr, *Chem. Eng. Sci.,* **50,** 2198, 1995a.

Ray, A. and R.W. Carr, *Chem. Eng. Sci.,* **50,** 3033, 1995b.

Ray, A., A.L. Tonkovich, R. Aris, and R.W. Carr, *Chem. Eng. Sci.,* **45,** 2431, 1990.

Ray, A., R.W. Carr, and R. Aris, *Chem. Eng. Sci.,* **49,** 469, 1994.

Ruthven, D.M., *Principles of Adsorption and Adsorption Processes,* Wiley & Sons, New York, 1984.

Takeuchi, K. and Y. Uraguchi, *J. Chem. Eng. Jpn.,* **9,** 164, 1976a.

Takeuchi, K. and Y. Uraguchi, *J. Chem. Eng. Jpn.,* **9,** 246, 1976b.

Takeuchi, K. and Y. Uraguchi, *J. Chem. Eng. Jpn.,* **10,** 72, 1977a.

Takeuchi, K. and Y. Uraguchi, *J. Chem. Eng. Jpn.,* **10,** 455, 1977b.

Tonkovich, A.L.Y. and R.W. Carr, *Chem. Eng. Sci.,* **49,** 4647, 1994.

Tonkovich, A.L.Y., R.W. Carr, and R. Aris, *Science,* **262,** 221, 1993.

Vaporciyan, G.G. and R.H. Kadlec, *AIChE J.,* **33,** 1334, 1987.

Vaporciyan, G.G. and R.H. Kadlec, *AIChE J.,* **35,** 831, 1989.

Viswanathan, S. and R. Aris, *Adv. Chem. Ser.,* **133,** 191, 1974a.

Viswanathan, S. and R. Aris, *SIAM-AMS Proc.,* **8,** 99, 1974b.

Wardwell, A.W., R.W. Carr, and R. Aris, in *Chemical Reaction Engineering,* J. Wei and C. Georgakis, Eds., ACS Symp. Ser., Boston, **196,** 1982, p. 297.

Westerterp, K.R., U.S. Patent 4,731,387, 1988.

Westerterp, K.R. and M. Kuczynski, *Chem. Eng. Sci.,* **42,** 1871, 1987.

Yang, R.T., *Gas Separation by Adsorption Processes,* Butterworths, Boston, 1987.

Chapter 6

REACTIVE MEMBRANE SEPARATION

José G. Sanchez Marcano and Theodore T. Tsotsis

6.1 Introduction

Membrane-based separation processes are finding today wide and ever-increasing use in the petrochemical, food, and pharmaceutical industries, in biotechnology, and in a variety of environmental applications, including the treatment of contaminated air and water streams. The most direct advantage of membrane separation processes, when compared to their more conventional counterparts (adsorption, absorption, distillation, etc.), is energy savings and reduction in the initial capital investment requirements. Membrane-based reactive separation processes, which seek to combine two distinct functions, i.e., reaction and separation, have been around as a concept since the early stages of the membrane field itself, but have only attracted substantial technical interest during the last decade or so (Hsieh, 1996). There is ongoing significant industrial interest in these processes because of their promise to be compact and less capital intensive, and their potential for substantial savings in the processing costs (Soria, 1995).

Membrane-based reactive separation processes (also known as catalytic membrane reactor processes) are attracting attention in catalytic reactor applications. In these reactor systems the membrane separation process is coupled with a catalytic reaction. When the separation and reaction processes are combined into a single unit, the membrane, besides providing the separation function, also often provides enhanced selectivity and/or yield. Membrane-based reactive separations were first utilized with reactions for which the continuous extraction of products would enhance the yield by shifting the equilibrium. Reactions of this type that have been investigated include dehydrogenation and esterification. Reactive separations also appear to be attractive for application to other types of reactions including hydrogenation and partial and total oxidation. In many reactor studies involving these reactions the use of membranes has been shown to increase the yield and selectivity. Published accounts of the application of membrane-based reactive separations in catalytic processes report the use of both catalytically active and inactive membranes of various types, shapes and configurations (Sanchez and Tsotsis, 1996). Reactor yield and reaction selectivity are found to be strongly dependent on the membrane characteristics, in addition to the more conventional process parameters. Reactor modeling has proven valuable for understanding the behavior of these systems (Tsotsis et al., 1993). It will continue to serve in the future as an important tool for predicting and optimizing the behavior, and for improving the efficiency of these processes. In this chapter, we will review first the broad spectrum of catalytic reactions and the different membrane types that have been utilized in reactive separation studies. Key aspects of process modeling, design, and optimization will then be discussed.

Biotechnology is another area in which membrane-based reactive separations are also attracting great interest. There, membrane processes are coupled with industrially important biological reactions. These include the broad class of fermentation-type processes, widely used in the biotechnology industry for the production of amino acids, antibiotics, and other fine chemicals. Membrane-based reactive separation processes are of interest here for the continuous elimination of metabolites necessary to maintain high reactor productivity. Membranes are also increasingly being utilized as hosts for the immobilization of bacteria, enzymes, or animal cells in the production of many high-value-added chemicals. Similar reactive separation processes are also finding application in the biological treatment of contaminated air and water streams. Many of these emerging applications will also be reviewed and evaluated in this chapter.

In the early stages of the membrane-based reactive separations field, the coupling of the two functions happened by simply connecting in series two physically distinct units, the reactor and membrane separator (see Figure 6.1(a)). The membrane reactor concept, shown in Figure 6.1(b), which combines two different processing units (a reactor and a membrane separator) into a single unit, was the result of natural process design evolution from the concept of Figure 6.1(a). There are obvious advantages resulting from the design configuration of Figure 6.1(b) relating to its compact design, and the capital and operating savings realized by the elimination of intermediate processing steps. Other advantages relate to the synergy that is being

(a)

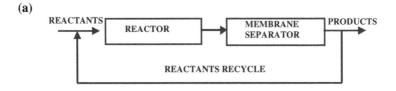

(b)

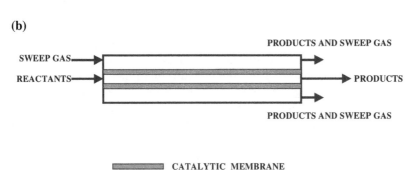

CATALYTIC MEMBRANE

Figure 6.1: (a) Conventional system, a reactor and a membrane separator; (b) catalytic membrane reactor.

realized between separation and reaction. This synergy is immediately obvious for reactions limited by thermodynamic equilibrium considerations, as is frequently the case with catalytic hydrocarbon dehydrogenation. There the continuous separation of one or more of the products (e.g., hydrogen) is reflected by an increase in yield and/or selectivity. Despite the obvious advantages of the reactor concept of Figure 6.1(b) over the design in Figure 6.1(a), the more conventional membrane-based reactive separation concept of Figure 6.1(a), because of its simplicity, is often the design of choice in biotechnological applications.

In the membrane reactor concept in Figure 6.1(b) the presence of the membrane (tubular, hollow fiber, or plate) helps to define two different chambers. These are the retentate chamber, where the reactants are fed and the reaction often takes place, and the permeate chamber. The latter is either swept by an inert gas or evacuated to maintain a differential pressure or concentration gradient for mass transfer between the two compartments.

Publications discussing the membrane reactor concept first started appearing in the late 1960s (Wood and Wise, 1968; Gryaznov et al., 1970). Most of the real progress in this area, however, has happened in the last 20 years (Hsieh, 1996). This parallels progress in the field of membrane processes. There the development of new membrane materials during the last 20 years has opened the world of membrane technology to a broader range of applications beyond the classical ones, which typically involve low-temperature microfiltration or ultrafiltration of liquids using polymeric membranes. The development of membranes made with a variety of inorganic materials has provided the opportunity to apply the catalytic membrane reactor concept for a much broader set of operating conditions. Inorganic membranes offer advantages in this regard over organic membranes, because they are stable at relatively high temperatures (> 373 K), and have good chemical and mechanical resistance (Soria, 1995). The progress, realized over the last 10 years, in the synthesis of stable microporous or dense inorganic materials for the preparation of membranes has been the key factor motivating the application of membrane-based reactive separations in the catalysis field. It should be noted, however, that recent developments in polymeric membranes (e.g., polyimides) are pushing the envelope for their application ($T \sim 300°C$). A number of studies, as a result, have also appeared discussing the use of such membranes in high-temperature membrane reactive applications (Rezac et al., 1994, 1995).

Many of the earlier applications involved equilibrium-limited reactions. In a rather short time, the field has progressed far beyond these applications. In some of the most recent studies, for example, the membrane separates intermediates and products from the reacting zone so that they do not deactivate the catalyst or undergo further undesirable reactions. In some of the other applications the membrane is not even required to be permselective; it only acts as a controlled reactive interface between reactants flowing on opposite sides of the membrane (Sloot et al., 1990; Torres et al., 1994). This significant progress in the field is reflected in the increasing number of publications on catalytic membrane reactors, which have grown exponentially over the last few years (Saracco et al., 1999). Detailed descriptions of the state of the art on this topic have been published along the way by Catalytica (1989), Armor (1989, 1992, 1995, 1998), Ilias and Govind (1989), Hsieh (1991, 1996), Zaspalis and Burggraaf (1991), Tsotsis et al. (1993), Zaman and Chakma (1994), Saracco and Specchia (1994, 1995), Saracco et al. (1994), Sanchez and Tsotsis (1996), and most recently Dixon (1999) and Saracco et al. (1999).

Inorganic membranes are also more frequently being utilized in biotechnology for the production of fine chemicals via the use of both enzyme and whole-cell bioreactors (Chang and Furusaki, 1991), and for large-scale environmental clean-up type applications (Brookes and Livingston, 1994). For these processes, which take place under milder conditions, organic membranes still remain the option of choice.

Both dense and porous membranes have found use in reactive separation applications. Dense membranes are made of polymers, metals and their alloys, and solid oxides. Dense polymeric membranes have not found extensive use in membrane-based reactive separations. For catalytic applications, except for some notable cases (Rezac et al., 1994, 1995), their use has been limited to low-temperature reactions (Feldman and Orchin, 1990; Kim and Datta, 1991; Troger et al., 1997), because they are perceived to have poor thermal resistance. Dense metal membranes consist mostly of noble metals, Pd, Pt, Ru, Rh, and Ir, Ag and their alloys, but most recently also of other hydrogen storage alloys, e.g., $LaNi_5$ (Uemura et al., 1998). These membranes (particularly Pd) have found extensive use in the early stages of the reactive separations field, starting as early as 1966, for the study of the ethane dehydrogenation reaction (Pfefferie, 1966). They were used extensively in the former Soviet Union by Gryaznov and co-workers (Gryaznov et al., 1986; Gryaznov and Smirnov, 1977; Gryaznov and Karavanov, 1979). These membranes have high selectivity towards hydrogen (Pd and its alloys) or oxygen (Ag) but their high cost, relatively low permeability, and limited availability have hampered their extensive industrial application. There is also concern over their mechanical properties, particularly embrittlement and fatigue as a result of repeated thermal cycling at high temperatures. There are ongoing significant efforts in this area for improving the characteristics of these membranes (Gryaznov, 1992). The approach here is to deposit thin metallic films on underlying porous substrates, aiming to improve permeance and cost without unduly impacting on selectivity (Bauxbaum and Kinney, 1996; Yeung and Varma, 1995; Yeung et al., 1995; Jeema et al., 1996; Jayaraman et al., 1995).

Another group of dense membranes, which has attracted significant attention in recent years (Bouwmeester and Burggraaf, 1996), consists of solid oxides (ZrO_2, Y_2O_3, Bi_2O_3) as well as solutions of mixed solid oxides (perovskites, brownmillerites, etc.). These materials act as solid electrolytes, allowing the transport of oxygen or hydrogen. They are finding use in catalytic membrane reactor applications involving partial and total oxidation reactions. They will be discussed further in this chapter.

Porous membranes, composed of glass and ceramic materials, as well as polymers, have also found use in membrane-based reactive separations. They are classified as macroporous (average pore diameter, $d_p > 50\,nm$), mesoporous ($50\,nm > d_p > 2\,nm$), or microporous ($d_p < 2\,nm$). The earlier studies in the area of membrane catalysis made use of commercially available mesoporous membranes (Vycor® glass, alumina, titania, or zirconia (Zaspalis and Burggraaf, 1991)). Recent efforts have focused on trying to increase the selectivity of these membranes towards small molecules. This has been accomplished either by decreasing the mean pore size and/or narrowing the pore size distribution of existing mesoporous membranes (Saracco et al., 1999) or by synthesizing microporous membranes from new materials altogether (e.g., zeolite membranes (van de Graaf et al., 1998)). Porous polymeric membranes have been the design choice in membrane bioreactors; they have also been recently used for low-temperature catalytic applications (Fritsch and Theis, 1997).

There is a multitude of different configurations that have been proposed in the literature in order to combine the membrane separation module and the catalytic reactor into a single unit. Sanchez and Tsotsis (1996) have classified these configurations into six basic types, as indicated in Table 6.1. This classification and acronyms will be used throughout this chapter and it will be useful to our readers to briefly review what the various configurations entail.

The most commonly utilized membrane reactor is the PBMR, in which the membrane provides only the separation function. The reaction function is provided (in catalytic applications) by a packed bed of catalyst particles placed in the interior or exterior membrane volumes. In the CMR configuration the membrane provides simultaneously the separation and reaction functions.

To accomplish this, one could use either an intrinsically catalytic membrane (e.g., zeolite or metallic membrane) or a membrane that has been made catalytic through activation by introducing catalytic sites by either impregnation or ion exchange. This process concept is finding wider acceptance in the membrane bioreactor area, rather than with the high-temperature catalytic reactors. In the latter case, the potential for the catalytic membrane to deactivate and, as a result, to require subsequent regeneration has made this technical concept less than attractive. On occasion, the membrane used in the PBMR configuration is also itself catalytically active, often unintentionally so (frequently the case for metallic membranes), but on occasion purposely so in order to provide an additional catalytic function. The membrane reactor in this case is named PBCMR. For better control of the process temperature, some authors have suggested that the packed bed

Table 6.1:   Classification of Membrane Reactors (Sanchez and Tsotsis, 1996).

Acronym	Description
CMR	Catalytic membrane reactor
CNMR	Catalytic nonpermselective membrane reactor
PBMR	Packed-bed membrane reactor
PBCMR	Packed-bed catalytic membrane reactor
FBMR	Fluidized-bed membrane reactor
FBCMR	Fluidized-bed catalytic membrane reactor

should be replaced by a fluidized bed (FBMR or FBCMR). In the CNMR configuration the membrane, typically, is not permselective. It is only used to provide a well-defined reactive interface.

Other more elaborate reactive separation configurations have also been reported in the technical literature. They include staged membrane reactors, membrane reactors with multiple feed ports, multitubular membrane reactors (Tecik et al., 1994; Omorjan et al., 1998), multiphase catalytic membrane reactors, etc. Considerable attention in many studies has been paid to the effect of various ways of operating these reactors, including, the means to minimize reactant loss (Wu and Liu, 1992; Tiscareno-Lechuga et al., 1996), the use of sweep gas under cocurrent or countercurrent operation (Itoh, 1995), or the use of a vacuum on the permeate side (see section on pervaporation membrane reactors).

6.2 Catalytic Reactive Separation Processes

6.2.1 Dehydrogenation Reactions

The catalytic dehydrogenation of light alkanes is, potentially, an important process for the production of alkenes, which are valuable starting chemical materials for a variety of applications. This reaction is endothermic and is, therefore, performed at relatively high temperatures, to improve the yield to alkenes, which is limited at lower temperatures by the thermodynamic equilibrium. Operation at high temperatures, however, results in catalyst deactivation (thus requiring frequent reactivation), and in the production of undesired byproducts. For these reasons this reaction has been from the beginning of the membrane reactor field the most obvious choice for the application of the catalytic membrane reactor concept, and one of the most commonly studied reaction systems.

As mentioned earlier, light alkane dehydrogenation was the reaction studied by Gryaznov and co-workers in their pioneering work (Gryaznov, 1986). In their dehydrogenation reaction studies, they used Pd or Pd alloy dense membranes, which were 100% selective towards hydrogen permeation. Comprehensive review papers on Pd membrane reactors have been published by the same group (Gryaznov, 1986), and also by Shu et al. (1991). In recent years, Gryaznov and co-workers have focused their attention on optimizing membrane reactor design in order to increase the available membrane area/reactor volume ratio, hoping to compensate for the relatively low membrane permeability. Despite the success in the laboratory of the application of Pd membranes to a variety of light alkane dehydrogenation reactions, their large-scale industrial application still remains problematic because of a number of well-documented problems (Saracco et al.,

1999). These problems include their high cost and limited commercial availability, questions concerning their mechanical and thermal stability, and in the case of alkane dehydrogenation, poisoning due to carbon deposition.

As one may surmise, the emphasis in recent years has been on improving the properties of these membranes. As noted previously, the approach taken here recently is to deposit thin films of metals like Pd and its alloys over a porous support (ceramic, Vycor® glass, or metal) using a variety of methods including electrodeless plating, sol–gel techniques, and deposition by magnetron sputtering. The aim of these efforts is to prepare a membrane with added mechanical strength due to the presence of the porous matrix, which maintains the high selectivity of the dense membrane, with improved hydrogen permeability due to the presence of a very thin (1–20 μm) metal film. Such membranes have been recently applied to butane and isobutane dehydrogenation. Matsuda et al. (1993), for example, obtained a considerable increase in isobutene yield (+600%) compared to the calculated equilibrium value at 673 K using a PBMR. The membrane was prepared by covering a porous alumina tube with a film of platinum by the electrodeless-plating technique. The ethane dehydrogenation reaction was studied by Gobina and Hughes (1994) using a porous Vycor® glass membrane modified with a thin film of Pd–Ag alloy deposited by a magnetron sputtering technique. They reported that the conversion increased from 2% (equilibrium value) to 18% at 660 K. In recent years dense polymeric membranes (Frisch et al., 1999) have also begun to be utilized in CMR dehydrogenation applications.

Reactive separations for light alkane dehydrogenation reactions to produce olefins have also been successfully performed using inorganic porous membranes. Champagnie et al. (1990, 1992) studied ethane dehydrogenation using a commercial 40 Å alumina membrane (Membralox®, U.S. Filter) impregnated with platinum as catalyst (CMR) or using the membrane with a packed bed of a commercial catalyst (PBMR or PBCMR). They reported increased yields in the conversion values attained in the absence of the membrane and greater than reference equilibrium conversions calculated at the tube and shell side pressures and temperatures. Porous membranes have also been applied to the propane dehydrogenation reaction to propylene, which is another very important raw material in the petrochemical industry. First mention of the potential application of porous membrane-based reactive separations to this reaction took place in 1988 in a British patent (Bitter, 1988), which proposed the use of γ-alumina membranes for the dehydrogenation of many organic compounds including propane. The first open literature application of these membranes for propane dehydrogenation is by Ziaka et al. (1993), who studied the reaction in a PBMR using a γ-alumina membrane and a commercial Pt/alumina catalyst. A number of other investigators have studied the reaction since then. The most recent investigation is by Weyten et al. (1997), who studied the propane dehydrogenation reaction in a PBMR using a commercial silica–alumina membrane with a good hydrogen permselectivity. They reported that the yield enhancement was, as expected, strongly dependent on the catalyst activity, membrane permeability, and space velocity.

As with metal membranes, the emphasis in recent years has been on improving the properties of the porous membranes. A class of membranes that have attracted attention are microporous silica membranes deposited by sol–gel (de Lange et al., 1995; Raman and Brinker, 1995; Ayral et al., 1995; de Voss and Verweij, 1998) or chemical vapor deposition (CVD) techniques (Tsapatsis and Gavalas, 1994; Kim and Gavalas, 1995; Morooka et al., 1995) on underlying mesoporous alumina or silica substrates. The use of a CVD silica membrane for the propane dehydrogenation reaction was previously described (Weyten et al., 1997). Ioannides and Gavalas (1993) utilized a dense silica membrane for isobutane dehydrogenation in a PBMR with modest increases in isobutene yield over those attained in a conventional reactor in the absence of the membrane. Silica membranes made by sol–gel techniques were also applied to propane dehydrogenation (Collins et al., 1996). Amorphous silica microporous membranes, unfortunately, have been shown to be sensitive to steam and coke in the presence of light alkanes and olefins under temperature conditions akin to catalytic dehydrogenation.

More promising for reactive separations involving gas-phase reactions appears to be the development and use in such applications of zeolite (and most recently of carbon molecular sieve (Itoh and Haraya, 1998)) membranes. One of the earliest mentions of the preparation of zeolite membranes is by Mobil workers (Haag and Tsikoyiannis, 1991), who reported in a U.S. patent the synthesis of very thin (and likely to be fragile) self-supported membranes made of ZSM-5 zeolite. The field of zeolite membrane synthesis has since then become a very active area (van de Graaf et al., 1998). A variety of apparently defect-free zeolite membranes have been prepared by depositing thin films of zeolite (silicalite and ZSM-5, SAPOs, etc.). Casanave et al. (1995) have reported the application of a silicalite membrane supported on a mesoporous alumina tube to the isobutane dehydrogenation reaction in a PBMR. An International Workshop on the Preparation of Zeolite Membranes and Films was held in June, 1998, in Gifu, Japan, during which a multitude of new and exciting developments were reported.

In addition to light alkanes, membrane-based reactive separations have also been applied to the dehydrogenation of a variety of other hydrocarbons. The most notable among these applications is the dehydrogenation of ethylbenzene to styrene, which is among the most important monomers used in the polymer industry (Gallaher et al., 1993; Tiscareno-Lechuga et al., 1993, 1996; Tagawa et al., 1998). A number of studies have reported the application of catalytic membrane reactors to this reaction using porous alumina and composite membranes, prepared by depositing thin films of Pd on porous ceramic or stainless steel substrates. In general, a PBMR has been used. The studies report an increase in the styrene yield compared to the yield attained in the absence of the membrane in a conventional packed-bed reactor. Tiscareno-Lechuga et al. (1993) also reported that the continuous removal of hydrogen by the membrane had the additional beneficial effect of slowing down the undesirable hydrodealkylation side reaction. The presence of the membrane thus resulted in an increase in the selectivity as well. A similar improvement in selectivity was reported by Raich and Foley (1998) during the dehydrogenation of ethanol to acetaldehyde in a PBMR utilizing a Pd membrane.

6.2.2 Hydrogenation Reactions

Membrane-based reactive separations have also been tested successfully for use in catalytic hydrogenation reactions. For hydrogenation reactions the membrane's role is in separating the liquid from the gaseous reactant (e.g., hydrogen) and providing a means for delivering this reactant at a controlled rate. Gryaznov et al. (1976) in the former Soviet Union were again among the first to report the application of catalytic membrane reactors to a hydrogenation reaction. They studied a number of hydrogenation reactions of value in the production of fine chemicals using Pd membranes. In such reactors hydrogen, which flows on one side of the membrane, diffuses selectively through the Pd membrane, and emerges on the other side in a highly reactive atomic form to react with the liquid substrate. To provide a source for the hydrogen, Gryaznov and co-workers proposed the coupling through the membrane of a dehydrogenation with a hydrogenation reaction. In this interesting concept the dehydrogenation reaction, occurring on one side of the membrane, produces the hydrogen which, after diffusing through the membrane, participates in the liquid phase hydrogenation reaction (Gryaznov, 1986; Gryaznov et al., 1976; Gryaznov and Karavanov, 1979). This technical approach also provides a means for coupling the reactions energetically, to attain an autothermal operation (several other novel concepts for providing the hydrogen have been proposed, including the recent interesting suggestion of directly integrating the reactor with an H_2O electrolysis unit (Itoh et al., 1998)). A number of industrial applications of the technology have been reported. They include the synthesis of vitamin K from quinone and acetic anhydride (Gryaznov and Smirnov, 1977), and the *cis*/*trans*-2-butene-1,4 diol hydrogenation to *cis*/*trans*-butanediol

(Gryaznov et al., 1983). More recently, the reactive separation concept has been applied (Ermilova et al., 1997) to cyclopentadiene hydrogenation utilizing a Pd–Ru ceramic–metallic composite membrane. In the area of dense membrane applications for hydrogenation reactions, a number of recent studies also report the use of proton-conducting solid oxide membranes (Otsuka and Yagi, 1994; Panagos et al., 1996). This is an exciting class of new materials with significant potential applications.

Hydrogenation reactions have also been studied with catalytic membrane reactors using porous membranes. In this case the membrane, in addition to being used as a contactor between the liquid and gaseous reactants, could potentially also act as a host for the catalyst, which is placed in the porous framework of the membrane. The first application was reported by Cini and Harold (1991) for the hydrogenation of α-methylstyrene to cumene. The authors report that this type of reactor presents the advantage of enhanced mass transfer rates between the reactant phases, and of a more efficient contact with the catalyst. A CMR was used by Torres et al. (1994) for nitrobenzene hydrogenation using a commercial Membralox® membrane, which was catalytically activated with Pt by ion exchange and impregnation. In this study the effects of the various operating reactor parameters were investigated in detail both experimentally and through the use of a theoretical model. Torres et al. (1994) showed that diffusional and kinetic resistances could be well controlled by adjusting the experimental parameters. More recently, Monticelli et al. (1997) reported the hydrogenation of cinnamaldehyde in a similar three-phase CMR. They concluded that with the CMR the selectivity towards hydrocinnamyl alcohol was better than the one attained with a classical slurry reactor. This result was explained by the absence of diffusional or other mass transport limitations when using the CMR. This is the result of the direct delivery of the gaseous reactant to the three-phase interface, located in a thin catalytic layer ($\sim 2\,\mu$m) at the external membrane surface.

6.2.3 Oxidation Reactions

Selective catalytic hydrocarbon oxidation reactions are difficult to implement because, in general, the intermediate products are more reactive towards oxygen than the original hydrocarbons. The net result is often total oxidation of the original substrate. One of the ways to increase the selectivity towards the intermediate products is to control the oxygen concentration along the reactor length. This can be conveniently implemented by means of the membrane reactor concept. The use of a membrane allows for the oxygen and the hydrocarbon reactants to be fed in different compartments. The most preferable configuration is the CMR, where the membrane itself is rendered catalytic, providing a reactive interface for the reaction to take place, while avoiding a long contact time between the desired products and oxygen. For the reasons previously outlined, however, the PBMR is the design of choice for most of the studies reported. With the use of the PBMR the oxygen concentration along the reactor length can be, within limits, carefully controlled to favorably influence the reactor selectivity. This, of course, is not possible in the case of the conventional fixed-bed reactors, where the oxygen concentration is maximum at the entrance and decreases monotonically along the length of the reactor. As a result the selectivity is typically low at the reactor inlet, where the reaction rate is the highest, which negatively impacts on the overall yield. One potential additional benefit of the application of the reactive separation concept to catalytic partial oxidation is that the separation of the oxidant and organic substrate creates reactor conditions less prone to explosions and other undesirable safety effects that are typically associated with the oxidation of gaseous hydrocarbons (Coronas et al., 1995a), thus potentially broadening the range of feasible operation (Dixon, 1999). Of concern are diminished reaction rates, due to the decreased oxygen partial pressures, and reactant hydrocarbon back-diffusion.

The use of membranes to implement, through controlled addition of oxygen, selective catalytic hydrocarbon oxidation has attracted considerable attention in recent years. The studies reported have made use of both dense and porous membranes. Dense membranes are made, typically, of metallic silver and its alloys, and various stabilized zirconias, as well as perovskites and brownmillerites. These materials are useful in preparing membranes because they are capable of transporting oxygen selectively. Porous membranes that have been utilized include zeolite and alumina either intact or impregnated by a variety of catalytic materials, including LaOCl, various perovskites, etc. Depending on their pore size and pore size distribution they have been used, with a varying degree of success, to maintain a controlled concentration of oxygen in the reaction side.

Silver and its alloys with other metals like vanadium are unique among metals in that they are very selective to oxygen transport. This phenomenon was first exploited by Gryaznov et al. (1986), who applied nonporous Ag membranes to the oxidation of ammonia, and to the oxidative dehydrogenation of ethanol to acetaldehyde. Most of the recent research efforts in the area of catalytic partial oxidation reactions have made use of solid oxide dense membranes. The earlier solid oxide dense membranes were conventional solid oxide electrolytes, most commonly titania-, yttria-, calcia-, or magnesia-stabilized zirconias (TiYSZ, YSZ, CSZ, MSZ) (Steele, 1992; Stoukides, 1988). These materials have good mobility of the oxygen anion in their lattice but lack sufficient electronic conductivity, thus necessitating the use of an external circuit. Significant advances in this area have been made by the introduction of perovskites (Teraoka et al., 1988) and brown-millerites (Schwartz et al., 1997; Sammels and Schwartz, 1998), which have both good ionic and electronic conductivity, and are also more thermally stable solids, better adapted to high-temperature applications. More novel membranes have also been prepared. For example, a U.S. patent (Mieville, 1994) reports the synthesis of membranes composed of mixed oxides and metals as selective oxygen transporters and catalysts for the methane oxidative coupling or partial hydrocarbon oxidation.

The direct conversion of methane into ethylene and ethane by oxidative coupling with oxygen is the prime example of a reaction using membrane reactors to which solid oxide membranes have been applied. This reaction has been extensively studied. TiYSZ, YSZ, CSZ, MSZ, doped Bi_2O_3, and oxygen and proton-conducting perovskite membranes have been used for methane oxidative coupling with or without the application of an external electrical potential (Otsuka et al., 1985; Hazburn, 1988; Nagamoto et al., 1990; Nozaki et al., 1992; Harold et al., 1994; Tsunoda et al., 1995; Hibino et al., 1995; Langguth et al., 1997; Zeng and Lin, 1997a,b, 1998). Though simulations (Wang and Lin, 1995) have predicted yields as high as 50%, in all the published studies the yield to ethane and ethylene has been reported to be less than 30%. This is also the maximum yield that has been obtained with a classical packed-bed reactor (Wolf, 1992). The problem with membrane reactors appears to be their low conversion, since generally the selectivity is reported to be higher than that of the conventional reactors (Bouwmeester and ten Elshof, 1997; Zeng and Lin, 1998). Androulakis and Reyes (1999) in a modeling study have shown that significant additional yield improvements could result by also simultaneously removing the C_2 product. A high-temperature membrane to accomplish this goal has yet to be developed (Dixon, 1999). A different approach to convert CH_4 into C_2 and higher hydrocarbons, utilizing a membrane reactor, was proposed by Garnier et al. (1996). They investigated a two-step process using a PBMR with a Pd–Ag membrane and a 5% Ru–Al_2O_3 catalyst. In the first step (which is favored at high temperatures) CH_4 is dehydrogenated into carbon and hydrogen. In the second step (which is favored at lower temperatures) the active carbon is rehydrogenated into C_{2+} hydrocarbons. The two-step, $CH_4 \rightarrow C_{2+}$ process had been studied earlier by a number of groups in a conventional reactor (Koerts et al., 1992). The advantage of the membrane reactor is that it lowers the temperature required in the first step to effect complete conversion of CH_4, thus favorably impacting on the economics. The robustness of the metal membrane in the reactive environment of step 1, and its ability to withstand thermal cycling between the two process steps, are the two key hurdles to be overcome. (Dehydrogenating ethane into hydrogen and carbon in a PBMR using a Pd–Ag membrane and graphite as catalyst with the goal of simply producing hydrogen was recently proposed by Murata et al. (1999). How one scales-up a process of this kind is somewhat unclear.)

Dense membranes have good permselectivity towards oxygen, which allows for the use of air during the partial oxidation reaction. On the other hand, their permeability is generally low, which places an upper limit on the reactor's efficiency. This, in turn, has motivated the use of porous membranes. The first studies, using porous membranes, were in the CMR configuration (Chanaud et al., 1995; Borges et al., 1995). They utilized catalytically active LaOCl membranes, which, unfortunately, were not very selective towards oxygen transport, and also did not exhibit the necessary stability at the reaction conditions. Efforts using the PBMR configuration have been more encouraging. The group at Zaragoza, Spain (Lafarga et al., 1994; Coronas et al., 1994a,b; Herguido et al., 1995), using nonpermselective, commercially available alumina membranes, and conventional oxidative coupling catalysts have attained promising yields ($\sim$25%). Ma et al. (1998) have reported recently a study of the oxidative coupling of methane using a lanthanum-stabilized porous γ-alumina membrane reactor in a PBMR configuration. The catalyst used was a packed bed of Mn–W–Na/SiO$_2$. The experimental results indicate an enhancement of the C$_2$ yield obtained with the PBMR configuration, when compared to the yield obtained with the more conventional co-feed configuration (Figure 6.2).

Another partial oxidation reaction that is attracting industrial attention for the application of reactive separations is the production of synthesis gas from methane. The earlier efforts made use of solid electrolytes. Eng and Stoukides (1988, 1991), for example, using a YSZ membrane in an electrochemical membrane reactor obtained a selectivity to CO and H$_2$ of up to 86%. Mazanec et al. (1992) integrated the exothermic CH$_4$ partial oxidation with endothermic CH$_4$ steam reforming to obtain high syngas yields (97%) in an electrochemical membrane reactor system that is potentially thermo-neutral. They also extended the natural gas upgrading to oxygen transport membranes that do not need external circuitry by the introduction of internally shorted, metal–YSZ composites and single-phase mixed conductors (Cable et al., 1990). This group recently reported (Mazanec, 1997) that their materials show stable microstructures after more than 500 hours of operation generating synthesis gas. Recent collaborative research efforts in this area by Amoco and the Argonne National Laboratory (Balachandran et al., 1995a,b, 1997, 1998) have made use of perovskite-like Sr(Co,Fe)O$_x$ materials. They have reported methane conversion efficiencies greater than 99% and stable membrane operation for over 900 hours. Workers at Eltron Research have made use of brownmillerite membranes (Schwartz et al.,

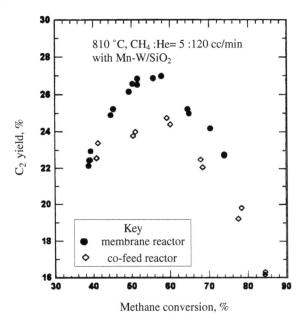

Figure 6.2: C_2 yields in a membrane and a co-feed reactor as function of methane conversion. (From Ma et al., 1998, with permission.)

1997; Sammels and Schwartz, 1998). They have also reported high yields to synthesis gas and stable operation for over one year (Sammels and Schwartz, 1998). Two industrial consortia were recently formed in the USA (Saracco et al., 1999; Mazanec, 1999) with the explicit purpose of commercializing synthesis gas production through the use of solid-oxide membrane-based reactor technology. One consortium, supported in part by the U.S. Department of Energy, is led by Air Products and Chemicals, and includes Babcock and Wilcox, Ceramatec, Eltron, ARCO, the Argonne National Laboratory, Penn State University, and the University of Pennsylvania. This group has announced plans for an eight-year, $85 million program. The other group is an industrial alliance among BP Amoco, Praxair, Statoil, SASOL, and Phillips Petroleum, which has also enlisted the help of a number of universities including MIT, and the Universities of Alaska-Fairbanks, Texas-Houston, Missouri-Rolla, and Illinois-Chicago to investigate fundamental material property issues. A third consortium with the same purpose seems to be in the works in Europe (Saracco et al., 1999). Despite these reported successes, concerns still appear to remain about the long-term stability of such materials in high-temperature reactive environments (Hendriksen et al., 1998). An effort to produce synthesis gas using a nonpermselective porous alumina membrane in a PBMR configuration is also under way (Yun et al., 1998).

A number of other partial hydrocarbon oxidation reactions have also been studied using catalytic membrane reactors. Using zirconia–calcia–alumina porous membranes, Zhong-Tao and Ru-Xuan (1994) studied the partial methane oxidation to methanol and formaldehyde. They report a selectivity of over 96% at 573 K. Oxidation of methane to formaldehyde using a zirconia–alumina macroporous membrane was studied by Yang et al. (1998). The membrane reactor gave better selectivities (for the same level of conversion) than the conventional packed-bed reactor, but the reported yields for both were rather small ($<1\%$). Direct catalytic oxidation of methane to methanol using catalytic membrane reactors was also reported by Lu et al. (1996) with rather modest improvements in yield. The use of membranes for improving the yield of the homogeneous partial oxidation of methane to methanol was reported by Liu et al. (1996). In their set-up, shown in Figure 6.3, the nonpermselective membrane provides a means for intimate contact for the reactants and a way to separate the external hot wall from an internal, water-cooled tube, whose purpose is to remove

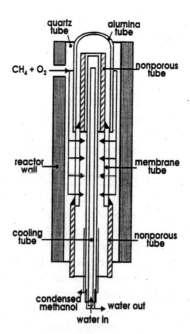

Figure 6.3: Schematic of the membrane reactor of Liu et al. (From Liu et al., 1996, with permission.)

CH_3OH by quenching. The concept of combining a membrane barrier with a quenching step (first proposed by Halloin and Wajc (1994) for the catalytic hydrogenation of toluene to methylcyclohexane) seemed to work well for methanol production, resulting in improved selectivity over the case in the absence of quenching. Hazburn (1988), in a U.S. patent, reported ethane and propane partial oxidation to ethylene and propylene oxides using TiYSZ membrane and a silver catalyst. Reactor yields as high as 7.5% were reported. Oxidation of ethylene to acetaldehyde has been studied using porous $V_2O_5/\gamma\text{-}Al_2O_3$ membranes (Harold et al., 1994). The selectivity towards the intermediate oxidation product reported in the CMR was higher than that of a conventional reactor. The selective oxidation of C_2–C_4 alkanes into oxygenates has been studied in an electrochemical membrane reactor using YSZ and ceria-based membranes by Hamakawa and co-workers (York et al., 1996; Hamakawa et al., 1997a, and references therein). Good selectivities were reported, but generally for very dilute hydrocarbon mixtures, and low conversions (the same group has also studied the production of synthesis gas from CH_4 using similar reactor system and membranes (Hamakawa et al., 1997b, and references therein)).

A number of studies using membrane reactors to control oxygen addition have investigated the oxidative dehydrogenation of low-molecular-weight hydrocarbons, including ethane to ethylene (Coronas et al., 1995b; Tonkovich et al., 1995, 1996), ethylene to acetylene (Keizer et al., 1994), propane to propylene (Pantazidis et al., 1995; Capannelli et al., 1996; Alfonso et al., 1999), and butane to butene (Tellez et al., 1997). Recently, the same concept has been applied to the oxidation of butane into maleic anhydride (Mallada et al., 1998), and ethylene epoxidation into ethylene oxide (Pena et al., 1998), where the potential advantage of the membrane reactor (thought to be intrinsically a safer device) would be to be able to operate with higher feed concentrations of the hydrocarbon. Generally, a beneficial effect on the selectivity was observed with the membrane reactor as compared with the classical packed-bed reactor. For the latter reaction Pena et al. (1998) observed that, in the conventional reactor, selectivity increases as the O_2/C_2H_4 molar feed ratio increases (in contrast to most other partial oxidation reactions). Therefore, it is more beneficial to use the membrane for distributing the C_2H_4 feed. The propane oxidative dimerization to benzene and hexadiene was studied by Azgui et al. (1994) in a CMR using a catalytic membrane made of bismuth–lanthanum oxide. A 36% reactor yield was obtained at 773 K.

A novel recent application involves the use of polymeric membranes, in which one incorporates active catalysts, as interfacial contactors to carry out partial oxidation (also hydrogenation and hydroformylation) of various organic compounds (Vankelecom et al., 1998). This concept is similar, in principle, to that previously applied to hydrogenation reactions. In addition to the potential advantages that were previously noted, in this case the membrane also allows carrying out the reaction without the need of a co-solvent, and prevents the loss of catalyst. The feasibility of the concept has been recently demonstrated by Wu et al. (1998) for the oxyfunctionalization of *n*-hexane to a mixture of hexanols and hexanones by hydrogen peroxide using titanium silicalite (TS1) as the catalyst embedded in a modified poly-dimethoxysiloxane (PDMS) membrane. Langhendries et al. (1998) have demonstrated the feasibility of the same concept with the oxidation of cyclohexane, cyclodecane, and *n*-dodecane using *t*-butylhydroperoxide as the oxidant and zeolite-encaged iron phthalocyanine as the catalyst embedded in a hydrophobic polymeric membrane matrix.

To summarize the above discussion, membrane-based reactive separations have been applied to partial oxidation reactions and some interesting results have been obtained. However, for the technical concept to have a significant beneficial impact, it will be necessary to develop mechanically and chemically stable membrane materials, which prevent the desired partially oxidized products from coming into prolonged contact with the oxidant. This is because, generally, the reactivity of the product towards oxygen or the oxide catalysts is higher than that of the hydrocarbon. Nonpermeable dense membranes are successful in that regard. However, the oxygen transport through the oxide lattice is low, when compared to its transport rate through porous materials. This requires the use of high temperatures of operation limiting the use of the membrane reactor concept to only a handful of reactions. Preparing dense, solid-oxide membranes with improved oxygen permeabilities is the subject of ongoing research efforts. An alternative approach is to synthesize porous ceramics with good oxygen permeability and selectivity. This is not a simple task and it is also the subject of considerable ongoing research. The development of microporous zeolite membranes capable of providing a high separation factor between oxygen and hydrocarbons offers hope in this area.

6.2.4   Other High-Temperature Applications

Ma et al. (1994) studied the H_2S decomposition in a PBMR configuration using a Vycor® porous glass membrane and a molybdenum sulfide catalyst. No sweep gas was used in the experiments to avoid the complications due to dilution effects, in determining the conversion enhancement. They concluded that in this case, in order to provide a significant shift in the reaction equilibrium, highly selective membranes may be required. Such a selective, composite Pt–V membrane was used for the study of this reaction by Edlund and Pledger (1994). Pd membranes, tested by the same group, had previously proved unstable in the same reaction environment. Pt–V membranes are, unfortunately, significantly less permeable than Pd membranes, and are, themselves, not free of some of the other problems associated with dense metallic membranes. The decomposition of dilute mixtures of NH_3 in a PBMR using Pd-alloy membranes was studied by Collins and Way (1994), and by Gobina et al. (1995). This application is of potential interest in the treatment of coal gasification streams, and the laboratory results showed promise. It would be interesting to see whether the same membranes prove robust in the real coal-gas environment. The use of a PBMR to study the hydrodechlorination of dichloroethane was recently reported by Chang et al. (1999). The reported potential advantage of the membrane would be in preferentially removing the byproduct HCl, which deactivates the catalyst. Chang et al. (1999) attribute the observed improved performance, however, to a dilution effect.

The coupling of membranes with a chemical reaction has also been applied to a number of other reactions of relevance to environmental applications. In contrast to the aforementioned two applications, for the applications discussed below the membrane simply plays the role of providing the reactive interface between reactants flowing on either membrane side (this is the case with some of the two-phase hydrogenation reactions previously discussed). For these reactions the goal is to attain full conversion of the pollutant and to avoid its slippage out of the reactor. The concept was first demonstrated by Zaspalis et al. (1991) for the selective catalytic reduction (SCR) of nitric oxide with ammonia to nitrogen and water. The role of the membrane in this case is to create a localized reaction interface, in which the nitric oxide completely reacts with ammonia. More recently, Saracco et al. (1999) extended the concept by depositing a catalyst in the pores of a fly-ash filter to combine NO SCR by NH_3 with particulate removal. Sloot et al. (1990, 1991) studied the Claus reaction between hydrogen sulfide and sulfur dioxide in a CMR configuration using a catalytically active, but nonpermselective membrane. They fed the hydrogen sulfide and sulfur dioxide on the opposite sides of the membrane and the reaction took place within the membrane itself. They demonstrated that if the reaction rate is sufficiently fast, a sharp reaction front is formed within the membrane, thus preventing slippage from either side. One then does not need to use a permselective membrane. They concluded, further, that this kind of system is very attractive, because by properly adjusting the operating conditions, undesired side reactions can be avoided. More recently, the same principle has been demonstrated with experiments and model calculations with CO (Veldsink et al., 1994) and H_2S oxidation (Neomagus et al., 1998). A different concept was recently suggested by Pina et al. (1996a,b) and Irusta et al. (1998). They have studied the use of CMR for destruction of volatile organic compounds

(VOC) in contaminated air streams. The idea is to create a more effective contact of the VOC with the catalyst by forcing the VOC-containing air stream through the membrane. Jacoby et al. (1998) have applied the same idea to the photocatalytic oxidation of carbonyl compounds and bioaerosols using a porous film of TiO_2 (acting both as an exclusion filter and a photo-catalytic layer) cast on an underlying support. A similar CMR concept was also applied by Binkerd et al. (1996) for the oxidative coupling of methane, and by Lange et al. (1998) and Lambert and Gonzalez (1999) for selective hydrogenations. For these reactions the benefits result from diminished side reactions between the desired intermediate products and the reactants (H_2, O_2), due to the short contact times.

Steam reforming of methane is one of the most interesting reactions, in which catalytic membrane reactors have been applied. Uemiya et al. (1991b) studied this reaction using a membrane prepared by coating a thin Pd film on a Vycor® glass tube utilizing an electrodeless plating technique. They observed significant enhancements in methane conversion over the calcu-lated equilibrium values. Interestingly, the effect of reactor pressure on the membrane reactor conversion is quite the opposite of what would be expected for the same reaction being carried out in a conventional reactor. According to Kikuchi et al. (1994) this behavior is indicative of the fact that in the membrane reactor the conversion is limited by the hydrogen diffusion through the metallic membrane. Kikuchi (1998) recently reported that these composite membranes are being evaluated by Tokyo Gas and Mitsubishi Heavy Industries in membrane reformers, in polymer electrolyte fuel cell systems. Adris et al. (1992, 1994), Adris and Grace (1997), and Mleczko et al. (1996) have studied the same reaction in an FBMR. This is a very inter-esting reactor design, which combines a catalytic fluidized bed with a bundle of palladium membranes. Performance was shown to be sensitive to the way the membrane area is distributed between the dense and dilute phases in the bed. The reaction has also been studied in a PBMR configuration utilizing mesoporous membranes but with somewhat more modest gains in conver-sion (Minet et al., 1992). The dry (i.e., using CO_2 rather than steam) reform-ing of methane is also recently attracting attention. The reaction has been studied in a PBMR configuration using both porous ceramic and dense Pd membranes (Raybold and Huff, 1999; Prabhu and Oyama, 1999). Despite the significant attention that methane reforming has received in the mem-brane-based reactive separations area, questions still remain about the potential effect that hydrogen removal through the membrane may have on accelerated catalyst deactivation due to coking and poisoning by H_2S (Hou and Hughes, 1998). The overall economic benefits, furthermore, still remain rather debatable, at least in the context of large-scale applications (Aasberg-Petersen et al., 1998; Hojlund Nielsen et al., 1998). Methanol reforming for the production of H_2 using a PBMR and Pd-type membranes has also been investigated recently by a number of groups (Lin et al., 1998; Hara et al., 1999) in the context of mobile fuel cell applications. Since membrane stability issues dictate operating at temperatures $> 300°C$ (Obradovic and Meldon, 1998), there is no beneficial effect due to equilibrium conversion displacement. Suggested potential benefits include reduced reactor volumes and a more compact unit design.

There have also been a number of membrane reactor applications to the Sabatier reaction, i.e., the catalytic hydrogenation of CO_2 to produce CH_4 and H_2O, of current interest for CO_2 capture and beneficiation. One of the earliest studies is by Gryaznov and co-workers (Gryaznov et al., 1981; Serov et al., 1983). They utilized a thin Pd–Ru foil as a hydrogen permselective membrane. They fed CO_2 on one side of the membrane and hydrogen on the other side. On the CO_2 membrane side they electrolytically deposited a porous film of Ni to act as the catalyst for the methanation reaction. In a recent study, Ohya et al. (1997) used a SiO_2 mesoporous membrane to carry out the reaction at somewhat elevated pressures (0.2 MPa). Higher pressures favor the equilibrium of this reaction. More importantly, for the reactive separation application, Ohya et al. (1997) envision that the SiO_2 membranes become water permselective under these conditions. In the experiments, H_2 and CO_2 mixtures (H_2/CO_2 feed ratio in the range 1–5) are fed on the one side of the membrane over a commercial 0.5% Ru on alumina catalyst. The water produced during the reaction permeates selectively through the membrane. Removing the product from the reaction zone increases the reactor conversion. In the range of the space velocities investigated (0.03–0.123 s^{-1}) and temperatures (480–719 K), an 18% maximum increase in conversion was observed over the reaction conversion attained in the absence of the membrane. An interesting application of membrane reactor technology to CO_2 capture was recently reported by Nishiguchi et al. (1998). In this application, CO_2 was catalytically hydrogenated to produce methane and water. The methane thus produced was then fed into a membrane reactor utilizing a Pd membrane, operating at 500°C, to be converted over a Ni/SiO_2 catalyst into graphitic carbon and hydrogen. In this two-stage reactor system over 70% of the CO_2 was thus converted into graphitic carbon. CO_2 hydrogenation to produce synthesis gas was studied by Uemura et al. (1998). In their reactor the membrane was made of a $LaNi_5$ alloy, typically utilized in hydrogen storage devices. The hydrogen was provided by the cyclohexane dehydrogenation reaction coupled to CO_2 hydrogenation through the membrane, which also acted as catalyst for both reactions. CO_2 hydrogenation to produce methanol in a membrane reactor at low temperatures ($< 200°C$) has been studied by Struis et al. (1996). Their study utilized Li-exchanged perfluorinated hydrocarbon membranes, which selectively permeated both products of the reaction (H_2O and CH_3OH). Polymeric membranes (PDMS, polyamide) impregnated with catalytically active components (active carbons, USY, and β-zeolites) were used by Vital et al. (1998) to study the catalytic hydration of α-pinene into monoterpenes, alcohols, and hydrocarbons. The use of the membranes appears to have a favorable influence on selectivity. Finally, in a recent study van de Graaf et al. (1999) reported the use of silicalite-1 membrane in a reactor to study the room temperature metathesis reaction of propene into ethylene and 2-butene, and the isomerization of *cis*-2-butene into *trans*-2-butene. The membrane at these conditions shows modestly preferential permeation of *trans*-2-butene over propene and *cis*-2-butene, and the use of the membrane results in noticeable improvements in conversion over the calculated equilibrium values.

6.2.5 Economic Considerations

Despite considerable technical attention, in recent years, application of membrane-based reactive separations to large-scale catalytic reaction processes is still lacking. Insight as to why this is so is offered in a published report by van Veen et al. (1996), who have investigated the feasibility of membrane reactor (MR) technology using porous membranes. They evaluated three reactions, namely propane dehydrogenation, ethylbenzene dehydrogenation to styrene, and the water–gas shift reaction (WGS) (the latter application is discussed further in the next section).

For propane dehydrogenation to propylene they compared the MR technology to two commercial catalytic dehydrogenation processes, the Catofin process (Lummus/Air Products) and the Oleflex process (UOP). (To put the whole discussion into a proper context, it should be noted that, for the most part, propylene is produced by steam cracker plants and refinery operations, and that catalytic technologies, in general, are not particularly attractive.) The first process diagram they investigated is a modification of the adiabatic Oleflex process, which consists of four reactors as shown in Figure 6.4. They considered four different process concepts: (1) a Knudsen membrane separator after the first, second, and third reactors; (2) a Knudsen membrane after

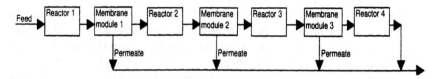

Figure 6.4: Process flow diagram including a membrane module after each reactor. (From van Veen et al., 1996, with permission.)

the third reactor only; (3) three ideal membrane separators (i.e., permeable only to hydrogen) after the first, second, and third reactors, which remove all the hydrogen formed in the reaction; (4) same as case 3 but with increased inlet temperature in the reactor.

Only cases 3 and 4 show any improvement in propylene yield over the base Oleflex process. The authors further concluded that moderate improvements in membrane permselectivity (say 10 for H_2 vs. C_3H_8) give marginal improvement in yield. Increasing the feed-side pressure, the amount of hydrogen in the feed and using a sweep gas also had questionable value. Only lowering the pressure of the permeate side had some noticeable effect on the yield.

They also evaluated isothermal MR concepts and compared them in performance with the adiabatic Catofin and Oleflex processes. They studied two different type processes using Knudsen diffusion membranes: a process called CMRL, patterned after the commercial Oleflex process, with low propane conversion, and a process called CMRH, patterned after the commercial Catofin process with high propane conversion. They have calculated the return on investment (ROI) for all four processes. Though marginally better than the commercial processes, the ROI for all four processes evaluated is not very attractive. A sensitivity analysis indicates that for the ROI of the MR processes to be attractive a price difference between propane and propylene of more that $300 per ton is required. Though published calculations have only been performed for the propane–propylene pair, it is not unreasonable to assume that similar conclusions apply to other alkane–alkene pairs (a number of other industrial companies have also carried out similar technical/economic evaluations (Ward et al., 1994)).

Van Veen et al. (1996) also evaluated the ethylbenzene to styrene reaction. The MR configurations studied (see Figure 6.5) incorporate the ceramic membrane within the reactor unit itself in a PBMR configuration. Three configurations have been examined. In configuration A, hydrogen permeates through the membrane and is carried away with steam. In configuration B, hydrogen is swept away and burned with air to provide heat for the endothermic reaction (a concept proposed earlier by Itoh and his co-workers (Itoh and Govind, 1989; Itoh, 1990)). This results in higher outlet temperatures, which results in higher conversions. The third configuration they examined, C, uses oxygen instead of hydrogen permeable membranes. Air is used as the source of oxygen and permeates through the membrane to carry out the oxidative dehydrogenation of ethylbenzene. From the preliminary evaluations, configurations B and C did not prove promising because of the loss of hydrogen, and potentially of styrene and ethylbenzene. Only configuration A was subsequently studied in greater detail. The conclusion was that the membrane reactors do not provide significant increases in the yield

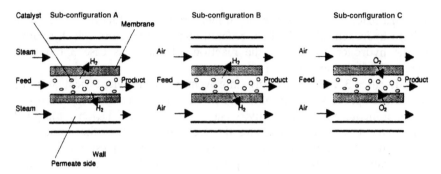

Figure 6.5: Membrane reactor sub-configurations. (From van Veen et al., 1996, with permission.)

(primarily due to slow reaction kinetics) to provide enough gain in the ROI to justify the significant investment required to put together even a retrofit plant. For the typical plant size considered by van Veen et al. (1996), the cost of the membranes alone was estimated to be ~$250 million. The economic evaluation for the water–gas shift reaction has shown it to be the most promising of all applications. The evaluation results are discussed further in the next section on emerging applications.

For the case of membrane bioreactors (for further technical details see the corresponding section below), the economics are reported to be more favorable (Tejayadi and Cheyran, 1995). The main reason for this is that the products are, in general, high value-added materials. Nevertheless, even for applications of this kind Tejayadi and Cheyran (1995) have reported that the membrane costs still represent up to 28% of the total capital investment and 5% of the operating costs. Lowering membrane costs still represents the arena where major improvements remain to be made.

6.2.6   Emerging Applications

Hydrogen production will certainly play a key role in the coal-based power plant of the future. All integrated gasification combined cycle (IGCC) plants are envisioned to produce hydrogen, either as a product for power production in a turbine or a fuel cell, or as a reactant to produce fuels and chemicals. Hydrogen-utilizing fuel cells, in particular, are significantly more efficient and produce lower SO_x and NO_x emissions than conventional combustion technology for power generation. For the coal-based power plant co-production concept to be successful it requires the development of better hydrogen separation technologies. Such technologies must have significantly improved efficiencies to impact on the plant economics, and must be robust to process conditions, which involve high temperatures and pressures, and the handling of complex mixtures, containing steam and reactive compounds like SO_2 and NH_3. High-temperature membrane-based technologies show the greatest promise for nonincremental technology leaps in this area.

Applying membrane-based reactive separation technology in the context of the IGCC concept has been discussed by a number of investigators. Pruschek et al. (1995), for example, have investigated the concept of CO_2 removal from coal-fired power stations. Their study indicates that net efficiencies of 40% are feasible. About 88% of the total CO_2 produced can be recovered using the IGCC concept employing a water–gas shift (WGS) reaction step and CO_2 separation either by physical absorption or, potentially, H_2/CO_2 separation with membranes.

The concept of combining the WGS reaction and the hydrogen separation step in a single unit has been investigated by a consortium of European academic and industrial entities (Bracht et al., 1997). They investigated the concept of the WGS membrane reactor (WGS-MR) for CO_2 removal in the IGCC system. In order to establish full insight into the possibilities of the application of such a reactor, a multidisciplinary feasibility study was carried out comprising system integration studies, catalyst research, membrane research, membrane reactor modeling, and bench-scale membrane reactor experiments (Xue et al., 1996). The study concluded that the application of the WGS-MR concept in IGCC systems is an attractive future option for CO_2 removal as compared to conventional options. The net efficiency of the IGCC process with integrated WGS-MR is 42.8% (LHV) with CO_2 recovery (80% based on coal input). This figure has to be compared with 46.7% (LHV) of an IGCC without CO_2 recovery and based on the same components, and with 40.5% (LHV) of an IGCC with conventional CO_2 removal. Moreover, an economic analysis indicates favorable investment and operational costs. The study concluded that development of the process is considered to be, in principle, technically feasible. However, the authors report that the currently available high-temperature selective gas separation membranes (noble metal and SiO_2) are not capable of withstanding the harsh environments found in the WGS reaction step and, therefore, further development in this area remains essential.

Though the idea by the European consortium to apply the WGS-MR concept to IGCC was novel, the WGS-MR concept itself was not. The WGS reaction, $CO + H_2O = CO_2 + H_2$, $(\Delta H^0 = -41.1\,\text{kJ mol}^{-1})$, is among the oldest catalytic reactions used to produce hydrogen for ammonia synthesis and numerous other applications. The reaction is exothermic, so lower temperatures favor higher conversions, but kinetic considerations favor higher temperatures. In practice, WGS systems consist of two reactors: a first reactor which is operated at high temperatures (623–673 K) using a high-temperature shift (HTS) catalyst, typically an Fe–Cr-based catalyst with high thermal stability; and a second reactor which is operating at low temperatures (ranging from 473 to 523 K), using an active Cu–Zn-based low-temperature shift (LTS) catalyst. To attain good conversions, typically H_2O/CO ratios significantly higher than 1 are utilized. The use of a membrane reactor (WGS-MR) provides the opportunity to attain high conversions in a single stage, with diminished needs for steam (see further discussion), and to simultaneously provide high-purity hydrogen. The first to study the WGS-MR concept was the group of Kikuchi and co-workers (Kikuchi et al., 1989; Uemiya et al., 1991a). They studied the WGS reaction at 673 K in a membrane reactor equipped with a Pd membrane deposited on a porous glass tube. They used a commercial Fe–Cr catalyst (Girdler G-3). Ar was used as the inert sweep gas and the reactor provided levels of conversion higher than equilibrium. Also, the amount of steam required to achieve reasonable levels of conversion was reduced.

Another emerging application of the membrane-assisted WGS concept (WGS-MR) is a process to recover tritium from tritiated water from breeder-blanket fluids in fusion reactor systems. A conceptual process model to accomplish this has been proposed by Violante et al. (1993). It uses two membrane reactor units. The first membrane reactor unit removes the hydrogen isotopes from the purge gas (He) via oxidation. The second unit uses the tritiated water to recover tritium using the WGS reaction. In a companion study, Basile et al. (1996) studied the WGS reaction in a membrane reactor using a composite Pd membrane created by a film of Pd deposited on the interior of a commercial alumina membrane by the so-called co-condensation technique. Their membrane was imperfect with an ideal H_2/N_2 separation factor of 8.2 at 595 K and 11.2 at 625 K. A Halder–Topsøe commercial catalyst was utilized. They determined the optimal conditions for reactor performance as a function of H_2O/CO ratio, temperature, reactor pressure, and gas feed flow rate with and without N_2 sweep gas. They observed that for the conventional reactor the maximum conversion attained increases with temperature. For the membrane reactor, on the other hand, an optimum temperature exists. Reaction studies at three temperatures (595, 615, and 633 K) show that reactor conversion increases with W/F. In a subsequent study, Tosti and Violante (1998) developed a model for the overall process, which accounts for the isotopic competition effect in the permeation process. The model, supported by a computer code, has been used to study the hydrogen isotopes extraction from the gas stream arising from a tritiated WGS reactor under conditions relevant to a solid breeding blanket of ITER (international thermonuclear experimental reactor) size.

As noted above, there is considerable interest today and significant emerging potential for the application of high-temperature membranes for both hydrogen and oxygen reactive separations. The main challenge that remains is in the development of membranes appropriate for reactive separation applications, under the harsh industrial temperature and pressure conditions. Though significant progress has been made recently to overcome some of these deficiencies, major breakthroughs in materials science are still needed before such membranes become viable in high-temperature/pressure separations of relevance to the aforementioned applications.

6.3 Pervaporation Membrane Reactors

Pervaporation membrane reactors are an emerging area of membrane-based reactive separations. An excellent review paper of the broader area of pervaporation-based, hybrid processes has been published recently (Lipnizki et al., 1999). The brief discussion here is an extract of the more comprehensive discussions presented in that paper, as well as in an earlier paper by our group (Zhu et al., 1996). Mostly nonbiological applications are discussed in this section. Some hybrid pervaporation–bioreactor applications are cited in the next section, and a more extensive discussion can be found in a recent publication (Lipnizki et al., 1998). Pervaporation (PV) has established itself, in recent years, as one of the most promising of membrane technologies. PV could potentially be useful in applications such as the dehydration of organic compounds, the removal/recovery of organic compounds from aqueous solutions, and the separation of organic mixtures. PV is often applied in combination with another technology as a hybrid process. Of these, PV-distillation and PV-reaction hybrid processes are already finding industrial applications. For PV-membrane-based reactive separations (PV-MR) the membrane either removes the desired product (mostly in biotechnological/wastewater applications, as described in the next section) or the undesired product (e.g., water for esterification reactions). Existing or proposed industrial processes for application of PV-MR include the production of ethyl and butyl acetate, ethyl and *n*-butyl oleate, diethyl tartrate, dimethyl urea, ethyl valerate, isopropyl and propyl propionate, and methyl isobutyl ketone, just to name a few.

As with the more conventional membrane-based reactive separations (PV-MR), applications range from the simpler designs, where the reactor and membrane are housed in separate units, to the more elaborate systems, for which the membrane and reaction functions are incorporated into the same unit. The principal class of reactions that have been studied by PV-MR are esterification reactions. Bitterlich et al. (1991), for example, studied a

PV-MR process for the production of butyl acetate. The conventional process uses mineral acid (H_2SO_4) as a catalyst. In the process, the acid must be neutralized with NaOH, before removing the water by distillation. In the process studied by Bitterlich et al. (1991), the catalytic action is provided by the acidic functionality of an anion exchange resin, thus eliminating the need for neutralization. The distillation column is replaced by a PV unit, resulting in energy savings. Dams and Krug (1991) analyzed various PV-MR designs using hydrophilic membranes. In two of the designs the PV unit was used to treat the vapor phase, while in the third design the PV unit directly treated the liquid phase from the esterification reactor. The third design attained high conversion. It required, however, an acid-resistant membrane (an analysis of this PV-MR design has been presented by David et al. (1991a,b) for the production of propyl propionate; a similar model was also developed by Keurentjes et al. (1994) for the production of diethyl tartrate). The other two designs (one involving a hybrid PV-distillation system, the other only a single PV unit) were not so successful in enhancing reactor conversion, but resulted, on the other hand, in overall higher energy savings. They could also utilize more conventional membranes. The PV-distillation hybrid system, in particular, is more amenable to plant expansion and retrofit. A PV-MR system utilizing an enzymatic esterification reaction (between erucic acid and hexadecyl alcohol) and an external PV unit employing hollow fiber membranes was studied by Nijhuis et al. (1992). The PV-MR attained over 90% conversions, while the esterification reactor alone was limited by equilibrium (~53% conversion). The process layout of a medium-sized esterification plant, used for the production of various esters, employing PV-MR systems was described by Bruschke et al. (1991). It consists of a cascade of two reactors, each individually coupled to two PV units. The construction of a larger esterification plant utilizing PV-MR technology was also mentioned in this report. In recent years, a number of groups have proposed the use of microporous inorganic membranes (SiO_2, zeolites, etc.) for PV applications. The use of a silica microporous membrane in a PV-MR system for the production of coating resins by a polycondensation reaction was recently reported by Bakker et al. (1998). The membrane showed good permselectivity to water coupled with a high permeance.

Some of the earlier reported applications, incorporating the membrane and reactor in the same unit, were by Kitta et al. (1988). They used a batch reactor, containing hydrophilic polyetherimide and polyimide membranes, to study the esterification of a number of carboxylic acids with ethanol. The presence of the membrane improved the reactor conversion beyond equilibrium, and thus reduced the reaction time and the amount of reactants needed to attain a target production rate for the ester. Hydrophilic polyimide membranes were also used by Ni et al. (1995) to study the esterification of valeric acid by ethanol in a PV-MR, with significant enhancements in conversion. Kwon et al. (1995) in a similar PV-MR system studied the production of butyl oleate by lipase-catalyzed esterification, using a hydrophilic cellulose acetate membrane. The PV-MR conversion was 92% as opposed to 61.1% without PV. A complete mathematical analysis for this type reactor was recently presented by Feng and Huang (1996). Tubular continuous PV-MRs have been studied by Waldburger et al. (1994) and by Zhu et al. (1996) for the production of ethyl acetate. Zhu et al. (1996) presented a mathematical model taking into account the thermodynamic nonidealities of the reactive liquid phase. Waldburger et al. (1994) discussed the economics of a number of process layouts involving the PV-MRs.

There are also a limited number of studies reporting on the application of PV-MR to reactions other than esterification. Heroin et al. (1991), for example, reported a pilot-plant-scale investigation, in which a PV unit employing a PVA-based hydrophilic membrane was coupled to a reactor producing dimethyl urea (DMU). DMU production results in an aqueous solution of CO_2 and methylamine. The use of the PV unit allowed the removal of most of the water, and direct recycle of the concentrated CO_2 and amine into the reactor, resulting in reduced energy costs and in a reduction in the amount of byproducts that must be disposed of. Matouq et al. (1994) have proposed the use of a PV membrane reactor incorporating a PVA membrane in the production of MTBE from methanol and *tert*-butyl alcohol. The PV-MR system is coupled with a more conventional reactive distillation unit. Combining two conventional batch reactors, with a PV unit (utilizing a PVA membrane) between them, has been suggested by Staudt-Bickel and Lichtenthaler (1996a,b) for the production of methyl isobutyl ketone (MIBK). The PV unit is used to dehydrate the product mixture from the first reactor. The membrane retentate, containing less than 0.1 wt% water, was fed for further processing into the second reactor.

These examples have shown that PV-MR separation processes are attracting significant attention and that the technology has found industrial applications. This is an area where significant activity is already under way, and many more advances are expected in the future.

6.4 Membrane-Based Reactive Separations for Biological Processes

The development of biological membrane-based reactive separation processes is a natural outcome of the extensive utilization that membranes find in the food industry. The dairy industry has been a pioneer in the use of membrane processes, which find application in the concentration of proteins from milk, or in the separation of valuable fractions of whey, including lactose, proteins, minerals, and various fats. These membrane-based processes (microfiltration, ultrafiltration, nanofiltration, and reverse osmosis) are typically performed under mild conditions (Mulder, 1996). These mild process conditions have made practical the use of membranes made from a variety of materials, including polymers and ceramics. Membranes are finding increased use in bioengineering. This is because biological systems produce, in general, a complex mixture of products. Only a limited number of such products are of value. The others have to be separated either during or after the reaction, to improve productivity, when these products turn out to inhibit biological function, or to achieve the desired product purity. Membranes offer a more biologically friendly, simpler, and less energy-intensive alternative to traditional separation processes. Another important area in which membranes have found a niche application is in the separation of biocatalysts, including enzymes and whole cells used for biochemical synthesis. These materials are very costly. Therefore, there is incentive to separate them from the product effluent, and to recycle them into the biological reactor. This is of particular importance when the metabolites of the bacteria or cells happen to be toxic to the microorganisms, or in the case of enzymatic reactions for which some of the reaction products inhibit the reaction. The net outcome of this action is to increase cellular concentration, to prolong biocatalyst life, and to effectively increase the product turnover rate. For example, during alcoholic fermentation by *S. cerevisiae* carried out in a membrane bioreactor, a productivity 30 times higher than that of a conventional batch reactor has been reported (Cheyran and Mehaia, 1984). Coupling ultrafiltration membranes with a bioreactor in the same pilot unit was the first membrane-based reactive separation that was described in the literature (Michaels, 1968; Flaschel et al., 1983).

Biochemical synthesis, utilizing both whole-cell and enzymatic reactors (Engasser, 1988; Hanemaaijer et al., 1987), is being used today for the production of a whole range of products, ranging from food and liquid fuels (ethanol), to plant metabolites, flavors, fragrances, and a variety of fine chemicals. Biochemical synthetic processes are of particular importance in the pharmaceutical industry, because they allow for the production of very complex molecules, like hormones, which cannot be produced safely and efficiently with other more conventional techniques. This is an area where membrane-based reactive separations are finding fertile grounds for application. Biological membrane reactors have been realized in many configurations. The simplest and most popular one consists of two separate but coupled units, one for the bioreactor and the other for the membrane module. This configuration has been used extensively, because of concerns with operation and process control of biological reactors, especially those involving fermentation or whole-cell conversions. The opinion here is that membrane-based separation processes, in which membrane and reaction functions are integrated into a single unit, are more difficult to control than their simpler counterparts, where the coupled membrane and reaction functions are housed in separate units (Hanemaaijer et al., 1987).

The most common application of this simple reactive separation application uses the membrane to separate the products or metabolites while retaining the enzymes or cells, which are then continuously recycled back to the bioreactor (Figure 6.6). Other applications involve the efficient delivery of one of the reactants (e.g., bubble-free aeration) using, for example, a microporous silicone membrane tube (Rissom et al., 1997).

There are a number of published studies in which the reaction and separation functions are integrated into the same unit. For enzymatic conversions, in particular, one often utilizes a hollow-fiber reactor, where enzymes are immobilized in the porous part of the hollow fiber membrane. This reactor is very similar in function to the CMR that was described previously for catalytic reactions. One important advantage that this type of membrane reactor offers over the more conventional enzymatic conversion bioreactors is the longer contact times of the reactants with the enzymes, due to the high membrane-surface/reactor-volume ratio. Low reactant residence times are a problem with conventional reactors and good efficiency is typically obtained only with rapid reactions (Marc et al., 1982). Hollow-fiber membrane bioreactors are compact systems with high productivity. The downside of this type of reactor is that cleaning of the hollow-fiber membranes is difficult, especially when the biocatalysts are whole cells or bacteria that are immobilized in the small pores of the fibers. To avoid this difficulty, PBMR configurations have also been tested, in which the biocatalysts instead are immobilized in porous beads, creating a packed porous bed, thus simplifying the membrane cleaning action.

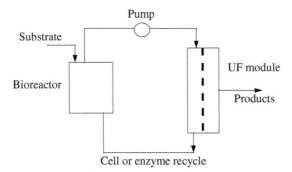

Figure 6.6: Bioreactor system coupled with an ultrafiltration (UF) unit.

There are many important reactions for which membrane-based reactive separations are finding application. The process is well suited, for example, for the enzymatic transformation (hydrolysis) of macromolecules (for more detailed reviews see, for example, Prazeres and Cabral, 1994; Cheyran and Mehaia, 1986). It has been extensively applied for protein hydrolysis and for enzymatic reactions, which require co-factor recycling (Engasser, 1988; Wandrey, 1987). One of the earliest successes realized with a membrane bioreactor was L-phenylalanine synthesis from acetamidocinnamic acid utilizing NADH (β-nicotinamide adenine dinucleotide) as a co-factor (Schmidt et al., 1987). In this study, the co-factor was grafted onto a soluble polymer, to increase its molecular size and to make it possible to recycle by ultrafiltration (UF). The combination of a conventional bioreactor with a UF membrane has been utilized in a number of other important synthesis reactions. An example is growth hormone biosynthesis obtained by the bacteria *E. coli* (Legoux et al., 1987, 1990). Another example is the synthesis of homochiral cyanohydrins (Bauer et al., 1993). Paolucci-Jeanjean et al. (1998) have reported the production of low-molecular-weight hydrolysates from the reaction of cassava starch over α-amylase. In this case the UF membrane separates the enzyme and substrate from the reaction products. Good enzyme recycle (without noticeable leaks) and productivity were obtained. Houng et al. (1992) had similar good success with maltose hydrolysis using the same type of reactor system. Clarification of fruit juice is another interesting application of membrane-based reactive separations. In the more conventional process after the enzymatic reaction pulp treatment step, a filtration step over diatomaceous earth is carried out. This traditional filtration-type process produces a lot of solid waste and results in costly enzyme loss.

This is a process that can be significantly improved upon by either using combined membrane filtration for enzyme recovery and recycle into a bioreactor or a more compact CMR-type system with the biocatalyst immobilized on the membrane itself. Figure 6.7 shows the proposed production process scheme.

Membrane-based reactive separations are also very interesting, not only for enzymatic transformations, but also for biological synthesis by whole cells. Crespo et al. (1992), for example, studied the production of propionic and lactic acids by two different bacteria in a membrane bioreactor. They concluded that a high cell concentration could be sustained in the membrane bioreactor for high production of metabolites. They observed, however, a problem that is frequently reported with membrane bioreactors, namely a decrease in the permeate flow rate with time on stream. This could be the result of biofouling (see further discussion below) or simply a manifestation of changes in viscosity and other fluid properties

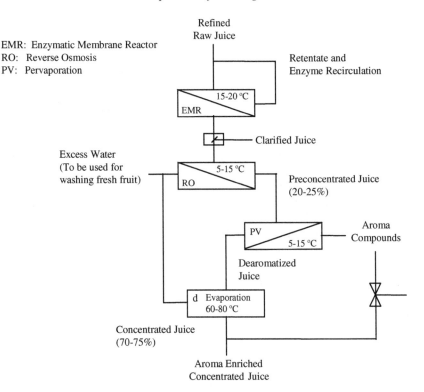

Figure 6.7: Enzymatic membrane reactor process for apple juice clarification. (From Cuperus et al., 1997, with permission.)

with time, as biomass accumulates in the reactor. In the latter case, the productivity could still be negatively impacted by a decrease in the heat and mass transfer (e.g., oxygen transfer for aerobic reactions). This problem can, at least in principle, be avoided by maintaining the cellular concentration below a certain critical level (Xavier et al., 1995). Koyama et al. (1987) have reported the synthesis of L-aspartic acid by *E. coli* in a bioreactor equipped with ultrafiltration hollow fibers. One of the most interesting reactor configurations utilized consisted of the cells growing in the space in between the fibers, with the metabolites continuously extracted through the hollow fibers. This reactor mimics the behavior of organs of living organisms, which also make use of a system of small capillaries. For mammalian cells this type of membrane reactor has been shown to result in high productivity (Hopkinson, 1985), and has been used for the synthesis of antibodies (Feder and Tolber, 1985).

Moueddeb et al. (1996) studied the transformation of lactose into lactic acid in a tubular membrane reactor, which contains two coaxial porous alumina tubes (Figure 6.8). The inside membrane contains a thin α-alumina microfiltration layer on its external wall, and the outside membrane has the same thin microfiltration layer on its external face. The bacteria are fixed onto the support macroporosity and, because of the presence of the microfiltration layers, are confined in the annular space defined by the separation walls. As noted previously, one of the most common problems encountered with membrane bioreactors is the phenomenon of biofouling, which is typically manifested by dramatic decreases in the permeate flow. This problem typically results because of the adsorption on the membrane surface and in its internal porous structure of various metabolites, that living bacteria and cells produce, and of the coagulated proteins from lysed cells (Figure 6.9). Biofouling is a difficult problem. If no remedial action is taken, the biomass continues to grow and, eventually, total membrane plugging will occur.

The conventional technique for dealing with the problem of membrane biofouling is to stop the reaction, after it has run for a predetermined period of time, and to carry out membrane cleaning and reactor sterilization operations. The critical factor for membrane bioreactor success is the use of appropriate membranes and cleaning techniques for each type of reaction,

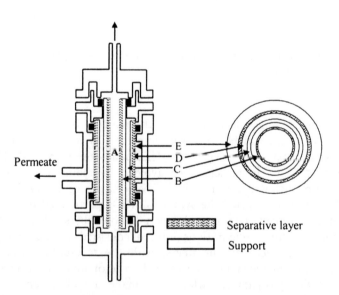

Figure 6.8: Schematic representation of the membrane bioreactor (axial and radial profiles). A, inner compartment; B and D, membranes; C, annular space; E, outer compartment. (From Moueddeb et al., 1996, with permission.)

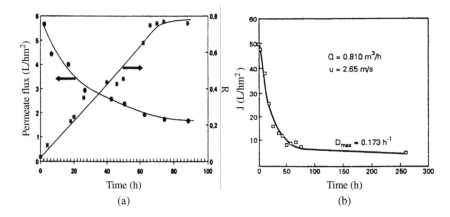

Figure 6.9: Permeate flux evolution of two different membrane bioreactor processes: (a) evolution of permeate flux for the lactic acid synthesis from lactose by *Lactobacillus rhamnosus* (Moueddeb et al., 1996, with permission); (b) evolution of permeate flux during propionic acid fermentation by *Propionibacterium acidi-propionici* (Crespo et al., 1992, with permission).

in order to avoid irreversible plugging. Because of the increased downtimes, due to the need to clean the membrane, it is necessary to maintain as high productivity as is possible during the running time. In recent years, a number of imaginative ideas have been proposed for in-situ clean-up of membranes. Significant improvements in performance have been reported for a number of cases (Fonade and Jaffrin, 1998).

6.5 Environmental Applications of Membrane Bioreactors

Another important and growing application of membrane bioreactors is in wastewater treatment. Conventional treatment of wastewater is often carried out by aerobic or anaerobic biological processes. These processes transform the complex organic contaminants typically found in the wastewater into simpler and harmless gaseous or water-soluble metabolites, together with some residual sludge. This type of conventional biological treatment has the disadvantage that one must, at some point, physically separate the biocatalyst from the treated water. Immobilization of the biomass on porous supports (conventional or trickling biofilters) (Devinny et al., 1998) alleviates the problem somewhat. For heavily polluted wastewaters, however, fast biomass growth and accumulation clogs the beds, and results in bed shutdowns and the need for frequent regeneration.

Membrane-based bioreactor processes present an alternative, attractive solution to the problem of biomass separation from the wastewater to be treated, since the membranes provide an effective barrier for microbes and other particles. The use of the membrane, furthermore, provides for more effective process control, since one can independently adjust the residence time in the fermentation vessel and the permeation rate through the membrane. One important advantage of the membrane-based bioreactor process is in the reduction of the size of the industrial unit. The coupling of the biological reaction and membrane separation function into the same unit, in general, requires a smaller volume when compared with the more conventional system consisting of a separate reactor and decanter. This, often, is an important consideration in many European countries, and in Japan.

Wastewater treatment and purification using membrane bioreactors have been studied in a number of research centers for several years (Anderson et al., 1986; Langlais et al., 1993; Scott et al., 1998; Ragona and Hall, 1998; Wisniewski et al., 1999), particularly in Europe, for example, by Lyonnaise des Eaux in France for the biological denitrification of drinking water, and the anaerobic treatment of food industry wastewater (Chaize and Huyard, 1991). The development of industrial units has been slow, however. The most difficult problem hindering development of membrane bioreactors for wastewater treatment has been membrane biofouling, which manifests itself with a decrease in the membrane permeation rate, due to the adsorption of metabolites onto the membrane's surface and into its pore structure. The Memtec Company (Monk, 1993) was among the first to develop a commercial, industrial membrane bioreactor process for wastewater treatment called the Membio® process. However, it turned out that the membrane flux was not sufficiently large to make the process economic. Membrane flux is very important in this area, because the water price is generally very low and the volumes to be treated are enormous. The same problem has attracted the attention of a number of other industrial companies. Two of the most notable efforts include the Biosep® process developed by CGE in France (Coté et al., 1997; Praderie et al., 1998) and the Kubota Process developed by the Kubota Corporation in Japan (Ishida et al., 1993; Churcose, 1997). Both systems consist of a bioreactor, in which a membrane in the form of a hollow fiber bundle is immersed in the activated sludge. Air, required by the aerobic biological process, is blown into the reactor through its bottom. This in turn assures agitation in the reactor vessel, which is important both in terms of improving reactor efficiency, and in terms of diminishing the biofouling problem. Plugging of the membranes was avoided by using membrane hollow fibers, whose dense permselective skin is on the outside part of the fiber, which is in contact with the suspension. The clean water is extracted through the membrane by applying vacuum in the inner side of the membranes. At least four such industrial units have been installed in France for industrial wastewater treatment, each with a capacity of up to 360 m^3 per day. The continuous operation of the system has proved that regular back-washing is no longer required. It also provides for a lower cost of operation than that of a conventional membrane system, while attaining remarkable reduction in floor space requirement.

6.6 Modeling of Membrane-Based Reactive Separation Processes

Dating back to the early studies by Gill et al. (1975), who developed models of tubular reactors with permeable walls, modeling and simulation of membrane reactors has been ongoing for quite a long time. Extensive discussion of the theoretical studies on catalytic membrane reactors has been previously presented by Tsotsis et al. (1993) and by Sanchez and Tsotsis (1996), and in a more recent work by Dixon (1999). The discussion here, by necessity a short summary, is patterned after these studies, which our readers who have further interest in the area should consult.

For reactors using mesoporous Knudsen-type membranes, Tsotsis et al. (1993) developed a general model for the PBCMR reactor, shown schematically in Figure 6.10, which accounts for volume change due to the reaction, nonisothermal effects, and pressure drops through the packed-bed regions, using the empirical Ergun equation. This model can be conveniently adapted for dense metal and solid membranes by replacing the Knudsen model of transport by other laws (e.g., Sievert's law for metal membranes) used to describe transport through the dense membranes. The model has

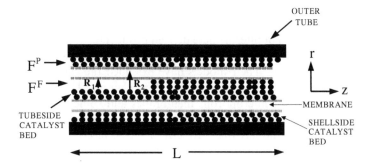

Figure 6.10: Schematic of membrane reactor for PBCMR model. (From Tsotsis et al., 1993, with permission.)

since been expanded in important ways by a number of other investigators. Dixon and co-workers (Becker et al., 1993) and Tayakout et al. (1995), for example, have presented models which, for the case of dilute reactive mixtures, account for the multilayer nature of most of the commercially available membranes. In contrast to the model of Tsotsis et al. (1993), which considers the membrane to consist of only one effective layer, these two groups write detailed diffusion equations in each individual membrane layer. The model of Tayakout et al., in addition, allows for the possibility of axial dispersion effects, potentially a matter of concern for laboratory reactors (Koukou et al., 1999). Tayakout et al. have applied their model to the cyclohexane dehydrogenation reaction, while Dixon and co-workers have modeled a variety of reaction systems including ethylbenzene dehydrogenation (Becker et al., 1993), methane partial oxidation using a dense solid-oxide membrane (Tsai et al., 1995), and the CO_2 and NO_x decomposition reactions (Dixon et al., 1994). A number of groups have concerned themselves with external mass transport effects, which could turn out to be a problem in large-scale operations (Papavassiliou et al., 1994). The effect of distributing the catalytic function nonuniformly in both membranes and catalysts has been investigated by Varma and co-workers (Yeung et al., 1994; Szegner et al., 1997). The all-important problem of how one compares performance, in terms of yield, between membrane and conventional reactors has been addressed by Reo et al. (1997a,b).

For macroporous membranes, but even on occasion for mesoporous membranes under high-pressure conditions (e.g., steam reforming), convective transport and molecular diffusion might be of concern. Van Swaaij and co-workers have developed models of membrane reactors utilizing the dusty gas model description of transport (Veldsink et al., 1998). They have applied these models to their CNMR studies of reactions, which require strict stoichiometric ratios, and to the combustion of CO and other hydrocarbons. Similar models have been utilized by Harold and co-workers (Harold and Lee, 1997) to model the application of membrane reactors, where the goal is to maximize the yield of intermediate desired product (e.g., partial oxidation reactions). Multiphase membrane reactors have been successfully modeled by Torres et al. (1994). Their models have extended the earlier models for such reactors by Akyurtlu et al. (1988) and by Cini and Harold (1991).

With the advent of microporous membranes and some early efforts for their utilization in membrane-based reactive applications, a need exists for developing effective models that properly describe molecular transport in such systems. For some of these membranes (zeolite, carbon), "average" pore structure dimensions are very close to the size of the transported molecules. Under such conditions the fluid is no longer a continuum, and the utility of models like the dusty gas model comes into question.

Nonequilibrium molecular dynamics techniques are finding increased utilization in this area (Xu et al., 1998; Sedigh et al., 1998).

Membrane bioreactors can, in principle, and have been modeled using approaches that have proven successful in the more conventional catalytic membrane reactor applications (Tsotsis et al., 1993). There are some additional complications, however, that typically relate to such reactors. To start, the rate equations associated with biological reactions are, in general, more complicated than those of reactions typically encountered in catalytic applications. For example, metabolites that are produced often inhibit the reaction. The reaction law to describe such an inhibition action is often very complex. As previously noted, biofilms that grow (either intentionally or unintentionally) on or within the membrane structure significantly affect the membrane flux, resulting on occasion in complete membrane plugging. Since most biological reactions take place in the liquid phase, one may be ill-advised to ignore the bulk-phase diffusional limitations. How one properly describes transport in the liquid phase, when biomass is present or within immobilized biofilms, is still a matter of ongoing research (Engasser, 1988). Some of the efforts to model membrane bioreactors seem not to have considered the complications that may result from the presence of the biomass. Tharakan and Chau (1987) developed a model and carried out numerical simulations to describe a radial-flow, hollow-fiber-membrane bioreactor, in which the biocatalyst consisted of a mammalian cell culture placed in the annular volume between the reactor cell and the hollow fibers. Their model utilizes the appropriate nonlinear kinetics to describe the substrate consumption; however, the flow patterns assumed for the model were based on those obtained with an empty reactor, and would be inappropriate when the annular volume is substantially filled with microorganisms (Figure 6.8). A similar type of model was developed by Cima et al. (1990), to describe a hollow-fiber perfusion system utilizing mouse adrenal tumor cells as a biocatalyst. In contrast, this model took into account the effect of the biomass that was present in the annular volume. A comprehensive model of a membrane bioreactor has been developed by Moueddeb et al. (1996). This model was intended to describe the reactor configuration referred to earlier in this chapter. The model equations were established taking into account the effect of the biomass in the annular volume and its effect on the permeate flow rate. The mass balance equations for the substrate (S) and the product (P) in cylindrical coordinates, utilized by Moueddeb et al. (1996) are:

$$D_1\left[\frac{\partial^2 S}{\partial r^2}+\frac{1}{r}\frac{\partial S}{\partial r}\right]-\frac{Q}{2\pi\varepsilon r L}\frac{\partial S}{\partial r}-R_S(S, X, P)=\frac{\partial S}{\partial t} \tag{6.1}$$

$$D_a\left[\frac{\partial^2 P}{\partial r^2}+\frac{1}{r}\frac{\partial P}{\partial r}\right]-\frac{Q}{2\pi\varepsilon rL}\frac{\partial P}{\partial r}+R_P(S,X,P)=\frac{\partial P}{\partial t} \tag{6.2}$$

In the above equations, D is the diffusion coefficient, Q is the permeate flux rate, and L is the reactor's length. The first term on the left side of each equation describes the radial dispersion and the second one describes the radial convection. Moueddeb et al. (1996) used in their model a biological rate kinetic expression (R_S or R_P), which was obtained by independent experiments and modeling in a batch reactor. They also made an effort to account for and correlate (Moueddeb, 1994) the decrease in the rate of permeate flow with the amount of produced biomass. The main result of this work was the theoretical determination of biomass concentration profiles, and the relationship between the rate of permeate flow and the calculated biomass concentration in the annular volume (Figure 6.11).

In conclusion, modeling is an important tool to provide insight into the complex reaction and mass and heat transport phenomena which take place in a membrane reactor. They provide guidance and help in terms of evaluating and optimizing the performance of such systems. The models that have been briefly reviewed here have served well the catalytic applications utilizing mesoporous and macroporous membranes. A message that emerges clearly from the application of these models is that for optimal operation of membrane reactors one must fine-tune and balance both the reactive and the separatory functions of the system. Too rapid a reaction rate or a not so permeable membrane will result in suboptimal performance (Raich and

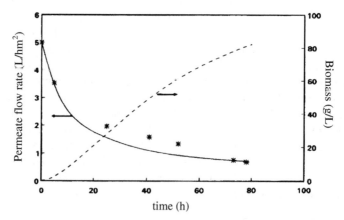

Figure 6.11: Evolution of permeate volumetric flow rate; calculated (———), experimental (*) and calculated microorganisms concentration profile in the annulus (- - - -). (From Moueddeb et al., 1996, with permission.)

Foley, 1995; Harold et al., 1994). In the modeling arena, real challenges still remain and significant advances must be made in the description of membrane bioreactors, and for catalytic applications involving microporous membranes (e.g., zeolites and carbon molecular sieve).

6.7 Conclusions

Membrane-based separation processes have been shown to be a promising option for the production of chemicals not only at the laboratory scale, but also in industrial processes. This type of process can be potentially applied to a broad spectrum of reactions ranging from the catalytic dehydrogenation of hydrocarbons, to hormone synthesis, and to the biological transformation of wastewaters. Membrane-based separation processes have already proven their economic and strategic value to low-temperature applications, like the bioproduction of fine, high-added-value chemicals, pervaporation, or wastewater treatment. This probably relates to the fact that such processes utilize polymeric and macroporous or mesoporous inorganic membranes, which are mature and readily available materials, and also have less severe requirements in terms of membrane housing and sealing. Good strides have also been made in recent years in a number of large-scale, high-temperature applications, a good example being synthesis gas or hydrogen production, through partial oxidation or steam reforming of light hydrocarbons. Despite the progress realized until now in this area, further advances must await the development of more stable and affordable membranes. The expected benefits of advances obtained in high-temperature membrane stability, and reduced costs are, of course, enormous given the size of the potential large-scale applications in the oil and petrochemical industry. For membrane-based reactive separations to make inroads into other areas, beyond the range of processes where they currently find application, careful process and reactor analysis must be carried out to evaluate the potential advantages over the conventional processes currently in place. For the petrochemical industry the task is compounded by the fact that the existing classical processes are mature and highly optimized, and the generally low process ROI does not provide much leeway in terms of initial capital investment. Membrane costs and availability are key factors here. As commented earlier, even for high-value-added biological products, membrane capital investment and operating costs are not negligible. For large-scale petrochemical applications such considerations remain probably the main factors hindering further progress.

Symbols

D_a	Lactic acid diffusion coefficient, $m^2\ s^{-1}$
D_l	Lactose diffusion coefficient, $m^2\ s^{-1}$
F	Molar flow rate, $mol\ s^{-1}$
L	Reactor length, m
P	Lactic acid concentration, $g\ L^{-1}$
$Q_{(r)}$	Permeate flux rate, $L\ h^{-1}$
r	Annulus radius, m
$R_{S,P}(S, X, P)$	Biological reaction kinetics, h^{-1}
S	Substrate or lactose concentration, $g\ L^{-1}$
t	Time, h
W	Catalyst weight, g
X	Bacteria concentration, $g\ L^{-1}$

Greek

ε	Porosity

References

Aasberg-Petersen, K., C.S. Nielsen, and S.L. Jorgensen, *Catal. Today*, **46**, 193, 1998.

Adris, A.M. and J.R. Grace, *Ind. Eng. Chem. Res.*, **36**, 45, 1997.

Adris, A.M., J.R. Grace, C.J. Lim, and S.S.E.H. Elnashaie, U.S. Patent Appl. 07965011; Canad. Patent Appl. 2081170, 1992.

Adris, A.M., C.J. Lim, and J.R. Grace, *Chem. Eng. Sci.*, **49**, 5833, 1994.

Akyurtlu, J.F., A. Akyurtlu, and C.E. Hamrin, Jr., *Chem. Eng. Commun.*, **66**, 169, 1988.

Alfonso, M.J., A. Julbe, D. Farrusseng, M. Menendez, and J. Santamaria, *Chem. Eng. Sci.*, **54**, 1265, 1999.

Anderson, G.K., C.B. Saw, and M.I.A.P. Fernandez, *Process. Biochem.*, **12**, 174, 1986.

Androulakis, I.P. and S.C. Reyes, *AIChE J.*, **45**, 860, 1999.

Armor, J.N., *Appl. Catal.*, **49**, 1, 1989.

Armor, J.N., *CHEMTECH*, **22**, 557, 1992.

Armor, J.N., *Catal. Today*, **25**, 199, 1995.

Armor, J.N., *J. Membr. Sci.*, **147**, 217, 1998.

Ayral, A., C. Balzer, T. Dabadie, C. Guizard, and A. Julbe, *Catal. Today*, **25**, 219, 1995.

Azgui, S., F. Guillaume, B. Taouk, and E. Bordes, *Proceedings of the 1st International Workshop on Catalytic Membranes*, Lyon-Villeurbanne, France, Sept. 26–28, 1994.

Bakker, W.J.W., I.A.A.C.M. Bos., W.L.P. Rutten, J.T.F. Keurentjes, and M. Wessling, "Application of Ceramic Pervaporation Membranes in

Polycondensation Reactions," in *Proceedings of the 5th International Conference on Inorganic Membranes*, Nagoya, Japan, June 22–26, 1998, p. 448.

Balachandran, U., J.T. Dusek, S.M. Sweeney, R.B. Poeppel, R.L. Mieville, P.S. Maiya, M.S. Kleefisch, S. Pei, T.P. Kobylinsky, and C.A. Udovich, *Ceram. Bull.*, **74**, 71, 1995a.

Balachandran, U., J.T. Dusek, R.L. Mieville, R.B. Poeppel, M.S. Kleefisch, S. Pei, T.P. Kobylinsky, C.A. Udovich, and A.C. Bose, *Appl. Catal., A*, **133**, 19, 1995b.

Balachandran, U., J.T. Dusek, P.S. Maiya, B. Ma, R.L. Mieville, M.S. Kleefisch, and C.A. Udovich, *Catal. Today*, **36**, 265, 1997.

Balachandran, U., P.S. Maiya, B. Ma, J.T. Dusek, R.L. Mieville, and J.J. Picciolo, "Development of Mixed-Conducting Ceramic Membranes for Converting Methane to Syngas," in *Proceedings of the 5th International Conference on Inorganic Membranes*, Nagoya, Japan, June 22–26, 1998.

Basile, A., A. Criscuoli, F. Santella, and E. Drioli, *Gas Sep. Pur.*, **10**, 243, 1996.

Bauer, B., H. Chmel, F. Effenberger, and H. Strathmann, *Proceedings of the International Congress on Membranes and Membrane Processes*, Heidelberg, Germany, August, 1993.

Bauxbaum, R.E. and A.B. Kinney, *Ind. Eng. Chem. Res.*, **35**, 530, 1996.

Becker, Y.L., A.G. Dixon, W.R. Moser, and Y.H. Ma, *J. Membrane Sci.*, **77**, 19, 1993.

Binkerd, C.R., Y.H. Ma, W.R. Moser, and A.G. Dixon, "An Experimental Study of the Oxidative Coupling of Methane in Porous Ceramic Radial-Flow Catalytic Membrane Reactors," in *Proceedings of the 4th International Conference on Inorganic Membranes*, D.E. Fain, Ed., Gatlinburg, TN, July 14–18, 1996, p. 441.

Bitter, J.G.A., Br. Patent 2,201,159, 1988.

Bitterlich, S., M. Meissner, and W. Hefner, *Proceedings of the 5th International Conference on Pervap. Proc. in the Chem. Ind.*, R. Brakish Ed., Brakish Material Corporation, Englewood, NJ, 273, 1991.

Borges, H., A.G. Fendler, C. Mirodatos, P. Chanaud, and A. Julbe, *Catal. Today*, **25**, 377, 1995.

Bouwmeester, H.J.M. and A.J. Burggraaf, *Fundamentals of Inorganic Membrane Science and Technology*, A.J. Burggraaf and L. Cot, Eds., Elsevier Science B.V., Amsterdam, The Netherlands, 435, 1996.

Bouwmeester, H.J.M. and J.E. ten Elshof, "Oxidative Coupling of Methane in a Dense Ceramic Membrane Reactor," in *Proceedings of the 4th Workshop: Optimisation of Catalytic Membrane Reactors Systems*, Oslo, Norway, May, 1997, p. 27.

Bracht, M., P.T. Alderliesten, R. Kloster, R. Pruschkek, G. Haupt, E. Xue, J.R. Ross, M.K. Koukou, and N. Papayannakos, *Energy Conversion and Management*, **38**, S159, 1997.

Brookes, P.R. and A.G. Livingston, *Water Res.*, **28**, 13, 1994.

Bruschke, H.E.A., G. Erlington, and W.H. Schneider, *Proceedings of the 5th International Conference on Pervap. Proc. in the Chem. Ind.*, R. Brakish, Ed., Brakish Material Corporation, Englewood, NJ, 310, 1991.

Cable, T.L., T.J. Mazanec, and J.G. Frye, Eur. Patent Appl. 0,399,833, Nov. 28, 1990.

Capannelli, G., E. Carosini, F. Cavani, O. Monticelli, and F. Trifiro, *Chem. Eng. Sci.*, **51**, 1817, 1996.

Casanave, D., A.G. Fendler, J. Sanchez, R. Loutaty, and J.A. Dalmon, *Catal. Today*, **25**, 309, 1995.

Catalytica Study Division, *Catalytica Study No. 4187*, Mountain View, CA, 1989.

Chaize, S. and A. Huyard, *Water Sci. Technol.*, **23**, 1591, 1991.

Champagnie, A.M., T.T. Tsotsis, R.G. Minet, and I.A. Webster, *Chem. Eng. Sci.*, **45**, 2423, 1990.

Champagnie, A.M., T.T. Tsotsis, R.G. Minet, and E. Wagner, *J. Catal.*, **134**, 713, 1992.

Chanaud, P., A. Julbe, A. Larbot, C. Guizard, L. Cot, H. Borges, A.G. Fendler, and C. Mirodatos, *Catal. Today*, **25**, 225, 1995.

Chang, C.C., C.M. Reo, and C.R.F. Lund, *Appl. Catal., B – Envir.*, **20**, 309, 1999.

Chang, H.N. and S. Furusaki, *Adv. Biochem. Eng. Bioeng.*, **44**, 27, 1991.

Cheyran, M. and M.A. Mehaia, *Proc. Biochem.*, **19**, 204, 1984.

Cheyran, M. and M.A. Mehaia, "Membrane Bioreactors," in *Membrane Separation in Biotechnology*, W.C. McGregor, Ed., Marcel Dekker, New York, 255, 1986.

Churcose, S., "Operating Experiences with the Kubota Submerged Membrane Activated Sludge Process," in *Proceedings of the 1st International Meeting on Membrane Bioreactors for Wastewater Treatment*, Cranfield University, March 5–6, 1997.

Cima, L.G., H.W. Blanch, and C.R. Wilke, *Bioprocess Eng.*, **5**, 19, 1990.

Cini, P. and M.P. Harold, *AIChE J.*, **37**, 997, 1991.

Collins, J.P. and J.D. Way, *J. Membr. Sci.*, **96**, 259, 1994.

Collins, J.P., R.W. Schwartz, R. Seghal, T.L. Ward, C.J. Brinker, G.P. Hagen, and C.A Udovich, *Ind. Eng. Chem. Res.*, **35**, 43, 1996.

Coronas, J., M. Menendez, and J. Santamaria, *Chem. Eng. Sci.*, **49**, 2015, 1994a.

Coronas, J., M. Menendez, and J. Santamaria, *Chem. Eng. Sci.*, **49**, 4749, 1994b.

Coronas, J., M. Menendez, and J. Santamaria, *J. Loss Prev. Process Ind.*, **8**, 97, 1995a.

Coronas, J., M. Menendez, and J. Santamaria, *Ind. Eng. Chem. Res.*, **34**, 4229, 1995b.

Coté, P., H. Buisson, C. Pound, and G. Araki, *Desalination*, **113**, 189, 1997.

Crespo, J.P.S.G., A.M.R.B. Xavier, M.T.O. Barreta, L.M.D. Goncalves, J.S. Almeida, and M.J.T. Carrondo, *Chem. Eng. Sci.*, **47**, 205, 1992.

Cuperus, F.P., S.Th. Bouwer, G. Boswinkel, R.W. van Gemert, and J.W. Veldsink, "The Upscaling of an Enzymatic Reactor for the Production of Apple Juice," in *Proceedings of the 4th Workshop: Optimisation of Catalytic Membrane Reactors Systems*, R. Bredesen, Ed., Oslo, Norway, May, 1997, p. 83.

Dams, A. and J. Krug, *Proceedings of the 5th International Conference on Pervap. Proc. in the Chem. Ind.*, R. Brakish, Ed., Brakish Material Corporation, Englewood, NJ, 338, 1991.

David, M.O., T.Q. Nguyen, and J. Neel, *Chem. Eng. Res. Des.*, **69**, 335, 1991a.

David, M.O., T.Q. Nguyen, and J. Neel, *Chem. Eng. Res. Des.*, **69**, 341, 1991b.

de Lange, R.S.A., J.H.A. Hekkink, K. Keizer, and A.J. Burggraaf, *J. Membr. Sci.*, **99**, 57, 1995.

Devinny, J.S., M.A. Deshusses, and T.S. Webster, *Biofiltration for Air Pollution Control*, CRC Press, Boca Raton, FL, 1998.

de Voss, R.M. and H. Verweij, *Science*, **279**, 1710, 1998.

Dixon, A. G., *Catalysis*, **14**, 40–92, 1999.

Dixon, A.G., W.R. Moser, and Y.H. Ma, *Ind. Eng. Chem. Res.*, **33**, 3015, 1994.

Edlund, D.J. and W.A. Pledger, *J. Membr. Sci.*, **94**, 111, 1994.

Eng, D. and M. Stoukides, "Partial Oxidation of Methane in a Solid Electrolyte Cell," in *Proceedings of the 9th International Congress on Catalysis*, Chem. Inst. of Canada, Ottawa, **2**, 974, 1988.

Eng, D. and M. Stoukides, *Catal. Rev. Sci. Eng.*, **33**, 375, 1991.

Engasser, J.M., "Reacteurs à Enzymes et Cellules Inmobilisées," in *Biotechnologie, Technique et Documentation*, Lavoisier, Paris, Chap. 4.2, 1988.

Ermilova, M.M., N.V. Orekhova, and V.M. Gryaznov, "Optimization of the Selective Hydrogenation Process by Membrane Catalysts," in *Proceedings of the 4th Workshop: Optimisation of Catalytic Membrane Reactors Systems*, R. Bredesen, Ed., Oslo, Norway, May, 1997, p. 187.

Feder, J. and R. Tolber, *Int. Biotechnol. Lab.*, **41**, June, 1985.

Feldman, J. and M. Orchin, *J. Catal.*, **63**, 213, 1990.

Feng, X. and R.Y.M. Huang, *Chem. Eng. Sci.*, **51**, 4673, 1996.

Flaschel, E., C. Wandrey, and M.R. Kula, *Adv. Biochem. Eng.*, **26**, 73, 1983.

Fonade, C. and M.Y. Jaffrin, "Régimes Instationnaires: Mise en Cuvre des Perturbations", in *Les Séparations par Membrane dans les Procédés de l'Industrie Alimentaire*, G. Dauphin, F. Rene, and P. Aimar, Eds., Technique et Documentation, Lavoisier, Paris, Chap. 4, 1998.

Frisch, H.L., S. Maaref, and H. Deng-Nemer, *J. Membr. Sci.*, **154**, 33, 1999.

Fritsch, D. and J. Theis, "First Results with Inorganic Modified Catalytic Polymer Membranes," in *Proceedings of the 4th Workshop: Optimisation of Catalytic Membrane Reactors Systems*, R. Bredesen, Ed., Oslo, Norway, May, 1997, p. 109.

Gallaher, Jr., G.R., T.E. Gerdes, and P.T.K. Liu, *Sep. Sci. Technol.*, **28**, 309, 1993.

Garnier, O., J. Shu, and B.P.A. Grandjean, "Catalytic Conversion of Methane into Hydrogen and Higher Hydrocarbons in a Pd-based Membrane Reactor," in *Proceedings of the 4th International Conference on Inorganic Membranes*, D.E. Fain, Ed., Gatlinburg, TN, July 14–18, 1996, p. 334.

Gill, W.N., E. Ruckenstein, and H.P. Hsieh, *Chem. Eng. Sci.*, **30**, 685, 1975.

Gobina, E. and R. Hughes, *J. Membr. Sci.*, **90**, 11, 1994.

Gobina, E.N., J.S. Oklany, and R. Hughes, *Ind. Eng. Chem. Res.*, **34**, 3777, 1995.

Gryaznov, V.M., *Platinum Met. Rev.*, **30**, 68, 1986.

Gryaznov, V.M., *Platinum Met. Rev.*, **36**, 70, 1992.

Gryaznov, V.M. and A.N. Karavanov, *Khim.- Farm. Zh.*, **13**, 74, 1979.

Gryaznov, V.M. and V.S. Smirnov, *Kinet. Catal.*, **18**, 485, 1977.

Gryaznov, V.M., V.S. Smirnov, L.K. Ivanova, and A.P. Mishchenko, *Dokl. Akad. Nauk. SSR*, **190**, 144, 1970.

Gryaznov, V.M., V.S. Smirnov, and M.G. Slin'ko, "Binary Palladium Alloys as Selective Membrane Catalysts," in *Proceedings of the 6th International Congress on Catalysis*, G.C. Bond, P.B. Wells, and F.C. Tompkins, Eds., **2**, 894, 1976.

Gryaznov, V.M., S.G. Gulyanova, Y.M. Serov, and V.D. Yagodovskii, *Zh. Fiz. Khim.*, **55**, 1306, 1981.

Gryaznov, V.M., A.N. Karavanov, T.M. Belosljudova, A.M. Ermolaev, A.P. Maganjunk, and I.K. Sarycheva, U.S. Patent 4,388,479, 1983.

Gryaznov, V.M., V.I. Vedernikov, and S.G. Gulyanova, *Kinet. Catal.*, **27**, 1,129, 1986.

Haag, W.O. and J.G. Tsikoyiannis, U.S. Patent 5,069,794, 1991.

Halloin, V.L. and S.J. Wajc, *Chem. Eng. Sci.*, **49**, 4691, 1994.

Hamakawa, S., T. Hayakawa, K. Suzuki, M. Shimizu, and K. Takehira, "Electrochemical Oxidation of Propene Using a Membrane Reactor with Solid Electrolyte," in *Proceedings of the 3rd World Congress on Oxidation Catalysis*, R.K. Grasseli, S.T. Oyama, A.M. Gaffney, and J.E. Lyons, Eds., 1323, 1997a.

Hamakawa, S., T. Hayakawa, K. Suzuki, R. Shiozaki, and K. Takehira, *Denki Kagaku*, **65**, 1049, 1997b.

Hanemaaijer, J.H., J. Stahouders, and S. Visser, "Fermentations and Enzyme Conversions in Membrane Reactors," in *Proceedings of the 4th European Congress in Biotechnology*, O.M. Neijssel, R.R. van der Meer, and K.Ch.A.M. Luyben, Eds., Elsevier, Amsterdam, The Netherlands, **1**, 119, 1987.

Hara, S., W.C. Xu, K. Sakaki, and N. Itoh, *Ind. Eng. Chem. Res.*, **38**, 488, 1999.

Harold, M.P. and C. Lee, *Chem. Eng. Sci.*, **52**, 1923, 1997.

Harold, M.P., C. Lee, A.J. Burggraaf, K. Keizer, V.T. Zaspalis, and R.S.A. de Lange, *Mater. Res. Soc. Bull.*, **34**, April, 1994.

Hazburn, E.A., U.S. Patent 4,791,079, 1988.

Hedriksen, P.V., P.H. Larsen, M. Mogensen, and F.W. Poulsen, "Prospects and Problems of Dense Oxygen Permeable Ceramic Membranes," in *III International Conference on Catalysis in Membrane Reactors*, Copenhagen, paper O8, Sept. 8–10, 1998.

Herguido, J., D. Lafarga, M. Menendez, J. Santamaria, and C. Guimon, *Catal. Today*, **25**, 263, 1995.

Heroin, C., W. Spiske and W. Hefner, "Dehydration in the Synthesis of Dimethylurea by Pervaporation," in *Proceedings of the 5th International Conference on Pervap. Proc. in the Chem. Ind.*, R. Brakish, Ed., Brakish Materials Corporation, Englewood, NJ, 1991, p. 349.

Hibino, T., T. Sato, K.I. Ushiki, and Y. Kuwahara, *J. Chem. Soc., Faraday Trans.*, **91**, 4419, 1995.

Hojlund-Nielsen, P.E., K. Aasberg-Petersen, and S. Laegsgaard-Jorgensen, "A Review on Hydrogen Selective Membranes—Obtained Results and Future

Perspectives," in *III International Conference on Catalysis in Membrane Reactors,* Copenhagen, paper O26, Sept. 8–10, 1998.

Hopkinson, J., *Biotechnology,* **3**, 225, 1985.

Hou, K. and R. Hughes, "The Effect of Hydrogen Removal on the Performance of a Membrane Reactor for Methane Steam Reforming," in *III International Conference on Catalysis in Membrane Reactors,* Copenhagen, paper O2, Sept. 8–10, 1998.

Houng, J.Y., J.Y. Chiou, and K.C. Chen, *Bioprocess Eng.,* **8**, 85, 1992.

Hsieh, H.P., *Catal. Rev. Sci. Eng.,* **33**, 1, 1991.

Hsieh, H.P., *Inorganic Membranes for Separation and Reaction,* Elsevier Science B.V., Amsterdam, The Netherlands, 1996.

Ilias, S. and R. Govind, *AIChE Symp. Ser.,* **85**, 268, 18, 1989.

Ioannides, T. and G.R. Gavalas, *J. Membr. Sci.,* **77**, 207, 1993.

Irusta, S., M.P. Pina, M. Menendez, and J. Santamaria, *Catal. Lett.,* **54,** 69, 1998.

Ishida, H., Y. Yamada, S. Matsumura, and M. Moro, "Application of Submerged Membrane Filter to Activated Sludge Process," in *Proceedings of the International Congress on Membrane and Membrane Processes,* R. Rautenbach, Ed., Heidelberg, Germany, Sept., 1993.

Itoh, N., *J. Chem. Eng. Japan,* **23**, 81, 1990.

Itoh, N., *Catal. Today,* **25**, 351, 1995.

Itoh, N. and R. Govind, *Ind. Eng. Chem. Res.,* **28**, 1557, 1989.

Itoh, N. and K. Haraya, "A Carbon Membrane Reactor," in *Proceedings of the 5th International Conference on Inorganic Membranes,* Nagoya, Japan, 1998, p. 338.

Itoh, N., S. Hara, K. Sakaki, T. Tsuji, and M. Hongo, "Electrochemical Coupling of Hydrocarbon Hydrogenation and Water Electrolysis," in *III International Conference on Catalysis in Membrane Reactors,* Copenhagen, paper O24, Sept. 8–10, 1998.

Jacoby, W.A., P.C. Maness, D.M. Blake, and E.J. Wolfrum, "Photocatalytic Membrane Reactor for Removal of Chemicals and Bioaerosols from Indoor Air," in *III International Conference on Catalysis in Membrane Reactors,* Copenhagen, paper O13, Sept. 8–10, 1998.

Jayaraman, V., Y.S. Lin, M. Pakala, and R.Y. Lin, *J. Membr. Sci.,* **99**, 89, 1995.

Jeema, N., J. Shu, S. Kaliaguine, and B.P.A. Grandjean, *Ind. Eng. Chem. Res.,* **35,** 973, 1996.

Keizer, K., V.T. Zaspalis, R.S.A. de Lange, M.P. Harold, and A.J. Burggraaf, *Membrane Processes in Separation and Purification,* J.G. Crespo and K.W. Boddeker, Eds., Kluwer Academic Publishers, The Netherlands, 415, 1994.

Keurentjes, J.T.F., G.H.R. Janssen, and J.J. Gorissen, *Chem. Eng. Sci.,* **49**, 4681, 1994.

Kikuchi, E., "Palladium Ceramic Membranes for Selective Hydrogen Permeation and Their Application to Membrane Reactor," in *Proceedings of the 1st International Workshop on Catalytic Membranes,* Lyon-Villeurbanne, France, Sept. 26–28, 1994.

Kikuchi, E., "Membrane Reactor Application to Hydrogen Production," in *III International Conference on Catalysis in Membrane Reactors*, Copenhagen, paper PL1, Sept. 8–10, 1998.

Kikuchi, E., S. Uemiya, N. Sato, H. Inoue, H. Ando, and T. Matsuda, *Chem. Lett.*, 489, 1989.

Kim, J. and R. Datta, *AIChE J.*, **37**, 1657, 1991.

Kim, S. and G.R. Gavalas, *Ind. Eng. Chem. Res.*, **34**, 168, 1995.

Kitta, N., S. Sasaki, K. Tanaka, K.I. Okamoto, and M. Yamamoto, *Chem. Lett.*, 2025, 1988.

Koerts, T., M.J.A.G. Deelen, and R.A. van Santen, *J. Catal.*, **138**, 101, 1992.

Koukou, M.K., N. Papayannakos, N.C. Markatos, M. Bracht, H.M. van Veen, and A. Roskam, *J. Membr. Sci.*, **155**, 241, 1999.

Koyama, Y., K. Shimazaki, K. Akashi, Y. Kawahara, K. Kubota, and H. Yoshii, "Production of L-Aspartic Acid by Membrane Reactor," in *Proceedings of the 4th European Congress in Biotechnology*, O.M. Neijssel, R.R. van der Meer, and K.Ch.A.M. Luyben, Eds., Elsevier, Amsterdam, The Netherlands, **1**, 119, 1987.

Kwon, S.J., K.M. Song, W.H. Hong, and J.S. Rhee, *Biotechnol. Bioeng.*, **46**, 393, 1995.

Lafarga, D., J. Santamaria, and M. Menendez, *Chem. Eng. Sci.*, **49**, 2005, 1994.

Lambert, C.K. and R.D. Gonzalez, *Catal. Lett.*, **57**, 1, 1999.

Lange, C., S. Storck, B. Tesche, and W.F. Maier, *J. Catal.*, **175**, 280, 1998.

Langguth, J., R. Dittmeyer, H. Hofmann, and G. Tomandl, *Appl. Catal., A.*, **158**, 287, 1997.

Langhendries, G., G.V. Baron, I.F.J. Vankelcom, R.F. Parton, and P.A. Jacobs, "Selective Hydrocarbon Oxidation Using Liquid-Phase Catalytic Membrane Reactors," in *III International Conference on Catalysis in Membrane Reactors*, Copenhagen, paper O10, Sept. 8–10, 1998.

Langlais, B., P. Denis, S. Triballeau, M. Faivre, and M.M. Bourbigot, *Water Sci. Technol.*, **25**, 219, 1993.

Legoux, R., P. Lepatois, J.E. Liauzun, B. Niaudet, W.G. Roskam, and W. Roskam, French Patent 2,597,114, 1987; U.S. Patent 4,945,047, 1990.

Lin, Y.M., G.L. Lee, and M.H. Rei, *Catal. Today*, **44**, 343, 1998.

Lipnizki, F., S. Hausmanns, G. Laufenberg, R. Field, and B. Kunz, *Chem. Eng. Technol.*, **70**, 1587, 1998.

Lipnizki, F., R.W. Field, and P.K. Ten, *J. Membr. Sci.*, **153**, 183, 1999.

Liu, Q., J. Rogut, B. Chen, J.L. Falconer, and R.D. Noble, *Fuel*, **75**, 1748, 1996.

Lu, G., S. Shen, and R. Wang, *Catal. Today*, **30**, 41, 1996.

Lu, Y., A. Ramachandran, Y.H. Ma, W.R. Moser, and A.G. Dixon, "Reactor Modeling of the Oxidative Coupling of Methane in Membrane Reactors," in *Proceedings of the 3rd International Conference on Inorganic Membranes*, Y.H. Ma, Ed., Worcester, MA, July 10–14, 1994, p. 657.

Ma, Y.H., W.R. Moser, S. Pien, and A.B. Shelekhin, "Experimental Study of H$_2$S Decomposition in a Membrane Reactor," in *Proceedings of the 3rd International Conference on Inorganic Membranes*, Y.H. Ma, Ed., Worcester, MA, July 10–14, 1994, p. 281.

Ma, Y.H., Y. Lu, A.G. Dixon, and W.R. Moser, "A Study of the Oxidative Coupling of Methane in a Lanthanum Stabilized Porous γ-Al$_2$O$_3$ Membrane Reactor," in *Proceedings of the 5th International Conference on Inorganic Membranes*, Nagoya, Japan, June 22–26, 1998, p. 330.

Mallada, R., M. Menendez, and J. Santamaria, "Synthesis of Maleic Anhydride in a Membrane Reactor Using a Butane Rich Feed," in *Proceedings of the 5th International Conference on Inorganic Membranes*, Nagoya, Japan, 1998, p. 612.

Marc, A., C. Burel, and J.M. Engasser, "Réacteurs Enzymatiques à Fibres Creuses pour L'Hydrolyse de l'Amidon et de Sacharose," in *Utilisation des Enzymes en Technologie Alimentaire*, P. Dupuy, Ed., Technique et Documentation, Lavoisier, Paris, 1982, p. 35.

Matouq, M., T. Tagawa, and S. Goto, *J. Chem. Eng. Japan*, **27**, 302, 1994.

Matsuda, T., I. Koike, N. Kubo, and E. Kikuchi, *Appl. Catal., A.*, **96**, 3, 1993.

Mazanec, T.J., "Electropox Gas Reforming," in *Proceedings of the 1st International Symposium on Ceramic Membranes*, H.U. Anderson, A.C. Khandkar, and M. Liu, Eds., **95-24**, 16, 1997.

Mazanec, T.J., Private communication, 1999.

Mazanec, T.J., T.L. Cable, and J.G. Frye, Jr., *Solid State Ionics*, **53**, 111, 1992.

Michaels, A.S., *Chem. Eng. Prog.*, **64**, 31, 1968.

Mieville, R.L., U.S. Patent 5,276,237, 1994.

Minet, R.G., S.P. Vasileiadis, and T.T. Tsotsis, "Experimental Studies of a Ceramic Membrane Reactor for the Steam/Methane Reaction at Moderate Temperatures, 400–700°C," in *Proceedings of the Symposium on Natural Gas Upgrading*, G.A. Huff and D.A. Scarpiello, Eds., **37**, 245, 1992.

Mleczko, L., T. Ostrowski, and T. Wurzel, *Chem. Eng. Sci.*, **51**, 3187, 1996.

Monk, C., "Application of Membranes in Waste Water Treatment," in *Proceedings of the Filtech Conference*, Publ. by The Filtration Society, October, 1993, p. 249.

Monticelli, O., A. Bezzi, A. Bottino, G. Capannelli, and A. Servida, "Hydrogenation of Cinnamaldehyde: the Use of Three Phase Catalytic Membrane Reactors," in *Proceedings of the 4th Workshop: Optimisation of Catalytic Membrane Reactors Systems*, Oslo, Norway, May, 1997, p. 109.

Morooka, S., S. Yan, K. Kusakabe, and Y. Akiyama, *J. Membr. Sci.*, **101**, 89, 1995.

Moueddeb, H., Ph.D. Thesis, UCB Lyon I.V., July, 1994.

Moueddeb, H., J. Sanchez, C. Bardot, and M. Fick, *J. Membr. Sci.*, **114**, 59, 1996.

Mulder, M., *Basic Principles of Membrane Technology*, Kluwar Academic Publishers, Dortrecht, 1996.

Murata, K., N. Ito, T. Hayakawa, K. Suzuki, and S. Hamakawa, *Chem. Commun.*, **7**, 573, 1999.

Nagamoto, H., K. Hayashi, and H. Inoue, *J. Catal.*, **126**, 671, 1990.

Neomagus, H.W.J.P., W.P.M. van Swaaij, and G.F. Versteeg, *J. Membr. Sci.*, **148**, 147, 1998.

Ni, X., Z. Xu, Y. Shi, and Y. Mu, *Water Treat.* **10**, 115, 1995.

Nijhuis, H.H., A. Kemperman, J.T.P. Derksen, and F.P. Cuperus, *Proceedings of the 6th International Conference on Pervap. Proc. in the Chem. Ind.*, R. Brakish, Ed., Brakish Material Corporation, Englewood, NJ, 368, 1992.

Nishiguchi, H., A. Fukunaga, Y. Miyashita, T. Ishihara, and Y. Takita, *Advances in Chemical Conversion for Mitigating Carbon Dioxide*, **114**, 147, 1998.

Nozaki, T., O. Yamazaki, K. Omata, and K. Fujimoto, *Chem. Eng. Sci.*, **47**, 2945, 1992.

Obradovic, B. and J.H. Meldon, "Hydrogen Permeability of Pd/Ag Membranes Under Methanol Reforming Conditions," in *Proceedings of the 10th Annual Meeting, North American Membrane Society*, Cleveland, May 16–20, 1998, p. 52.

Ohya, H., J. Fun, H. Kawamura, K. Itoh, H. Ohashi, M. Aihara, S. Tanisho, and Y. Nagishi, *J. Membr. Sci.*, **131**, 237, 1997.

Omorjan, R.P., R.N. Paunovic, and M.N. Tekic, *J. Membr. Sci.*, **138**, 57, 1998.

Otsuka, K. and T. Yagi, *J. Catal.*, **145**, 289, 1994.

Otsuka, K., S. Yokoyama, and A. Morikawa, *Chem. Lett.*, 319, 1985.

Panagos, E., L. Voudouris, and M. Stoukides, *Chem. Eng. Sci.*, **51**, 3175, 1996.

Pantazidis, A., J.A. Dalmon, and C. Mirodatos, *Catal. Today*, **25**, 403, 1995.

Paolucci-Jeanjean, D., M.P. Belleville, N. Zakhia, and G.M. Rios, *Proceedings of the 5th International Conference on Inorganic Membranes*, Nagoya, Japan, June 22–26, 1998, p. 146.

Papavassiliou, V.A., J.A. McHenry, E.W. Corcoran, H.W. Deckman, and J.H. Meldon, "High Flux Asymmetric Catalytic Membrane Reactor Optimization of Operating Conditions," in *Proceedings of the 1st International Workshop on Catalytic Membranes*, Lyon-Villeurbanne, France, September, 1994.

Pena, M., D.M. Carr, K.L. Yeung, and A. Varma, *Chem. Eng. Sci.,* **53**, 3821, 1998.

Pfefferie, W.C., U.S. Patent Appl. 3,290,406, 1966.

Pina, M.P., M. Menendez, and J. Santamaria, *Appl. Catal., B.,* **11**, L19, 1996a.

Pina, M.P., S. Irusta, M. Menendez, J. Santamaria, R. Hughes, and N. Boag, *Ind. Eng. Chem. Res.*, **36**, 4557, 1996b.

Prabhu, A.K. and S.T. Oyama, *Chem. Lett.,* **3**, 213, 1999.

Praderie, M., H. Buisson, H. Paillard, and T. Vouillon, *Galvano-organo-traitements de surface*, **685**, 390, 1998.

Prazeres, D.M.F. and J.M.S. Cabral, *Enzyme Microb. Technol.*, **16**, 738, 1994.

Pruschek, R., G. Oleljeklaus, V. Brand, G. Haupt, G. Zimmermann, and J.S. Ribberink, *Energy Conversion and Management*, **36**, 797, 1995.

Ragona, C.S.F. and E.R. Hall, *Water. Sci. Technol.*, **38**, 4, 1998.

Raich, B.A. and H.C. Foley, *Appl. Catal., A.,* **129**, 167, 1995.

Raich, B.A. and H.C. Foley, *Ind. Eng. Chem. Res.*, **37**, 3888, 1998.

Raman, N.K. and C.J. Brinker, *J. Membr. Sci.,* **105**, 273, 1995.

Raybold, T.M. and M.C. Huff, "CO_2 Reforming of CH_4 Over Supported Noble Metal Catalysts in a Pd Membrane Reactor," in *Proceedings of the 16th Meeting of the North American Catalysis Society*, Boston, MA, paper C-028, May 30–June 4, 1999.

Reo, C.M., L.A. Bernstein, and C.R.F. Lund, *AIChE J.,* **43**, 495, 1997a.

Reo, C.M., L.A. Bernstein, and C.R.F. Lund, *Chem. Eng. Sci.,* **52**, 3075, 1997b.

Rezac, M.E., W.J. Koros, and S.J. Miller, *J. Membr. Sci.,* **93**, 193, 1994.

Rezac, M.E., W.J. Koros, and S.J. Miller, *Ind. Eng. Chem. Res.*, **34**, 862, 1995.

Rissom, S., U. Schwarz-Linek, M. Vogel, V.I. Tishkov, and U. Kragl, *Tetrahedron: Asymmetry*, **8**, 2523, 1997.

Sammels, A.F. and M. Schwartz, "Catalytic Membrane Reactors for Spontaneous Synthesis Gas Production," in *III International Conference on Catalysis in Membrane Reactors*, Copenhagen, paper PL4, Sept. 8–10, 1998.

Sanchez, J. and T.T. Tsotsis, in *Fundamentals of Inorganic Membrane Science and Technology*, A.J. Burggraaf and L. Cot, Eds., Ch. 11, Elsevier, 1996.

Saracco, G. and V. Specchia, *Catal. Rev. Sci. Eng.*, **36**, 304, 1994.

Saracco, G. and V. Specchia, *Ind. Eng. Chem. Res.*, **34**, 1480, 1995.

Saracco, G. and V. Specchia, "Inorganic Membrane Reactors," in *Structured Catalysts and Reactors*, A. Cybulski and J.A. Moulijn, Eds., Marcel Dekker Inc., New York, 1998.

Saracco, G., G.F. Versteeg, and W.P.M. van Swaaij, *J. Membr. Sci.*, **95**, 105, 1994.

Saracco, G., H.W.J.P. Neomagus, G.F. Versteeg, and W.P.M. van Swaaij, *Chem. Eng. Sci.*, **54**, 1997, 1999.

Schmidt, E., W. Hummel, and C. Wandrey, "Continuous Production of L-Phenylalanine from Acetamidocinnamic Acid Using the Three-Enzyme-System Acylase/L-PHEDH/FDH," in *Proceedings of the 4th European Congress in Biotechnology*, O.M. Neijssel, R.R van der Meer, and K.Ch.A.M. Luyben, Eds., Elsevier, Amsterdam, The Netherlands, **2**, 189, 1987.

Schwartz, M., J.H. White, M.G. Myers, S. Deych, and A.F. Sammels, "The Use of Ceramic Membrane Reactors for the Partial Oxidation of Methane to Synthesis Gas," in *Preprints 213th ACS National Meeting,* San Francisco, CA, April 13–17, **42**, 1997.

Scott, J.A., D.J. Neilson, W. Liu, and P.N. Boon, *Water. Sci. Technol.*, **38**, 4-5, 1998.

Sedigh, M.G., W.J. Onstot, L. Xu, W.L. Peng, T.T. Tsotsis, and M. Sahimi, *J. Phys. Chem. A.*, **102**, 8580, 1998.

Serov, Y.M., V.M. Zhernosek, S.G. Gulyanova, and V.M. Gryaznov, *Kinet. and Catal.*, **24**, 303, 1983.

Shu, J., B.P.A. Grandjean, A. van Neste, and S. Kaliaguine, *Can. J. Chem. Eng.*, **69**, 1036, 1991.

Sloot, H.J., G.F. Versteeg, and W.P.M. van Swaaij, *Chem. Eng. Sci.*, **45**, 2415, 1990.

Sloot, H.J., C.A. Smolders, W.P.M. van Swaaij, and G.F. Versteeg, *AIChE J.*, **37**, 997, 1991.

Soria, R., *Catal. Today*, **25**, 285, 1995.

Staudt-Bickel, C.R. and N. Lichtenthaler, *J. Membr. Sci.*, **11**, 135, 1996a.

Staudt-Bickel, C.R. and N. Lichtenthaler, *Proceedings of the International Congress on Membranes and Membrane Processes,* 394, 1996b.

Steele, B.C.H., *Mater. Sci. Eng.*, **B13**, 79, 1992.

Stoukides, M., *Ind. Eng. Chem. Res.*, **27**, 1745, 1988.

Struis, R.P.W.J., S. Stucki, and M. Wiedorn, *J. Membr. Sci.*, **113**, 93, 1996.

Szegner, J., K.L. Yeung, and A. Varma, *AIChE J.*, **43**, 2059, 1997.

Tagawa, T., H. Itoh, and S. Goto, *Proceedings of the 5th International Conference on Inorganic Membranes,* Nagoya, Japan, June 22–26, 1998.

Tayakout, M., B. Bernauer, Y. Toure, and J. Sanchez, *J. Simul. Pract. and Theory*, **2**, 205 1995.

Tecik, M.N., R.N. Paunovic, and G.M. Ciric, *J. Membr. Sci.*, **96**, 213, 1994.

Tejayadi, S. and M. Cheyran, *Appl. Microbiol. Biotechnol.*, **43**, 242, 1995.

Tellez, C., M. Menendez, and J. Santamaria, *AIChE J.*, **43**, 777, 1997.

Teraoka, Y., H.M. Zhang, S. Furukawa, and N. Yamazoe, *Mater. Res. Bull.*, **33**, 51, 1988.

Tharakan, J.P. and P.C.K. Chau, *Bioeng.*, **29**, 657, 1987.

Tiscareno-Lechuga, F., C.G. Hill, Jr., and M.A. Anderson, *Appl. Catal. A.*, **96**, 33, 1993.

Tiscareno-Lechuga, F., C.G. Hill, Jr., and M.A. Anderson, *J. Membr. Sci.*, **118**, 65, 1996.

Tonkovich, A.L.Y., R. Secker, E. Reed, E. Roberts, and J. Cox, *Sep. Sci. Technol.*, **30**, 397, 1995.

Tonkovich, A.L.Y., J.L. Zilka, D.M. Jimenez, G.L. Roberts, and J.L. Cox, *Chem. Eng. Sci.*, **51**, 89, 1996.

Torres, M., J. Sanchez, J.A. Dalmon, B. Bernauer, and J. Lieto, *Ind. Eng. Chem. Res.*, **33**, 2421, 1994.

Tosti, S. and V. Violante, *Fusion Engineering and Design*, **43**, 93, 1998.

Troger, L., H. Hunnefeld, S. Nunes, M. Oehring, and D. Fritch, *J. Phys. Chem., B.*, **101**, 1279, 1997.

Tsai, C.Y., Y.H. Ma, W.R. Moser, and A.G. Dixon, *Chem. Eng. Commun.*, **134**, 107, 1995.

Tsapatsis, M. and G. Gavalas, *J. Membr. Sci.*, **87**, 281, 1994.

Tsotsis, T.T., R.G. Minet, A.M. Champagnie, and P.K.T. Liu, in *Computer Aided Design of Catalysts*, R. Becker and C. Pereira, Eds., Marcel Dekker, Inc., New York, 1993, p. 471.

Tsunoda, T., T. Hayakawa, Y. Imai, T. Kameyama, K. Takemira, and K. Fukuda, *Catal. Today*, **25**, 371, 1995.

Uemiya, S., N. Sato, H. Ando, and E. Kikuchi, *Ind. Eng. Chem. Res.*, **30**, 585, 1991a.

Uemiya, S., N. Sato, H. Ando, T. Matsuda, and E. Kikuchi, *Appl. Catal.*, **67**, 223, 1991b.

Uemura, Y., Y. Ohzuno and Y. Hatate, "A Membrane Reactor Using Hydrogen Storage Alloy for CO_2 Reduction," in *Proceedings of the 5th International Conference on Inorganic Membranes*, Nagoya, Japan, 1998, p. 620.

van de Graaf, J.N., F. Kapteijn, and J.A. Moulijn, "Catalytic Membranes," in *Structured Catalysts and Reactors*, A. Cybulski and J.A. Moulijn, Eds., Marcel Dekker Inc., New York, 1998.

van de Graaf, J.N., M. Zwiep, F. Kapteijn, and J.A. Moulijn, *Appl. Catal., A.*, **178**, 225, 1999.

Vankelecom, I.F.J., K. Vercruysse, P. Neys, D. Tas, K.B.M. Janssen, P. Knops-Gerrits, R.F. Parton, and P.A. Jacobs, "Dense Catalytic Membranes for Fine Chemicals Synthesis," in *III International Conference on Catalysis in Membrane Reactors*, Copenhagen, paper O12, Sept. 8–10, 1998.

van Veen, H.M., M. Bracht, E. Harmoen, and P.T. Alderliesten, in *Fundamentals of Inorganic Membrane Science and Technology*, Ch. 14, A.J. Burggraaf and L. Cot, Eds., Elsevier, 1996.

Veldsink, J.M., G.F. Versteeg, and W.P.M. van Swaaij, paper presented at the *1st International Workshop on Catalytic Membranes*, Lyon-Villeurbanne, France, Sept. 26–28, 1994.

Veldsink, J.W., R.M.J. van Damme, G.F. Versteeg, and W.P.M. van Swaaij, *Chem. Eng. Commun.*, **169**, 145, 1998.

Violante, V., A. Basile, and E. Drioli, *Fusion Engineering and Design*, **22**, 257, 1993.

Vital, A., A.M. Ramos, I.F. Silva, and H. Valente, "Hydration of α-pinene over Zeolites and Activated Carbon Dispersed in Polymeric Membranes," in *III International Conference on Catalysis in Membrane Reactors*, Copenhagen, paper P9, Sept. 8–10, 1998.

Waldburger, R., F. Widmer, and W. Heinzelmann, *Chem. Eng. Technol.*, **66**, 850, 1994.

Wandrey, C., "Fine Chemicals," in *Proceedings of the 4th European Congress in Biotechnology*, O.M. Neijssel, R.R. van der Meer, and K.Ch.A.M. Luyben, Eds., Elsevier, Amsterdam, The Netherlands, **3**, 22, 1987.

Wang, W. and Y.S. Lin, *J. Membr. Sci.*, **103,** 219, 1995.

Ward, T.L., G.P. Hagen, and C.A. Udovich, "Assessment of Inorganic Membrane Technology for Petrochemical Applications", in *Proceedings of the 3rd International Conference on Inorganic Membranes*, Y.H. Ma, Ed., Worcester, MA, July 10–14, 1994, p. 335.

Weyten, H., K. Keizer, A. Kinoo, J. Luyten, and R. Leysen, *AIChE J.*, **43,** 1819, 1997.

Wisniewski, C., A. Leon-Cruz, and A. Grasmick, *Biochem. Eng. J.,* **3**, 61, 1999.

Wolf, E.E., *Methane Conversion by Oxidative Processes*, Reinhold, New York, 1992.

Wood, B.J. and H. Wise, *J. Catal.*, **11**, 30, 1968.

Wu, J.S. and P.K.T. Liu, *Ind. Eng. Chem. Res.*, **31**, 322, 1992.

Wu, S., C. Bouchard, and S. Kaliaguine, "Zeolite Containing Catalytic Membranes as Interface Contactors," in *III International Conference on Catalysis in Membrane Reactors*, Copenhagen, paper O9, Sept. 8–10, 1998.

Xavier, A.M.R.B., L.M.D. Gonalves, J.L. Moreira, and M.J.T. Carrondo, *Biotechnol. Bioeng.*, **45**, 320, 1995.

Xu, L., M.G. Sedigh, M. Sahimi, and T.T. Tsotsis, *Phys. Rev. Lett.*, **80**, 3511, 1998.

Xue, E., M.O. Keefe, and J.R.H. Ross, *Catal. Today*, **30**, 107, 1996.

Yang, C., N. Xu, and J. Shi, "Numerical Simulation and Experimental Study of Catalytic Inert Membrane Reactor for Partial Oxidation of CH_4 to HCHO," in *Proceedings of the 5th International Conference on Inorganic Membranes*, Nagoya, Japan, 1998, p. 616.

Yeung, K.L. and A. Varma, *AIChE J.,* **41**, 4823, 1995.

Yeung, K.L., A. Aravind, R.J.X. Zawada, J. Szegner, G. Gao, and A.Varma, *Chem. Eng. Sci.*, **49**, 4823, 1994.

Yeung, K.L., J.M. Sebastian, and A. Varma, *Catal. Today*, **25**, 232, 1995.

York, A.P.E., S. Hamakawa, T. Hayakawa, K. Sato, T. Tsunoda, and K. Takehira, *J. Chem. Soc., Faraday Trans.*, **92**, 3579, 1996.

Yun, J., H. Hahm, M. Alibrando, and E.E. Wolf, "Methane Partial Oxidation in a Fast Flow Non-Selective Membrane Reactor," in *III International Conference on Catalysis in Membrane Reactors*, Copenhagen, paper O4, Sept. 8–10, 1998.

Zaman, J. and A. Chakma, *J. Membr. Sci.*, **92**, 1, 1994.

Zaspalis, V.T. and J. Burggraaf, in *Inorganic Membranes: Synthesis, Characteristics and Applications*, Ch. 7, R.R. Bhave, Ed., Reinhold, New York, 1991.

Zaspalis, V.T., W. van Praag, K. Keizer, J.G. van Ommen, J.R.H. Ross, and A.J. Burggraaf, *Appl. Catal.*, **74**, 249, 1991.

Zeng, Y. and Y.S. Lin, *Ind. Eng. Chem. Res.*, **36**, 277, 1997a.

Zeng, Y. and Y.S. Lin, *Appl. Catal., A.*, **159**, 101, 1997b.

Zeng, Y. and Y.S. Lin, "Experimental Study of Oxidative Coupling of Methane in Dense Ceramic Membrane Reactor," in *Proceedings of the 5th International Conference on Inorganic Membranes*, Nagoya, Japan, June 22–26, 1998, p. 354.

Zhong-Tao, H. and L. Ru-Xuan, paper presented at the *1st International Workshop on Catalytic Membranes*, Lyon-Villeurbanne, France, Sept. 26–28, 1994.

Zhu, Y., R.G. Minet, and T.T. Tsotsis, *Chem. Eng. Sci.*, **51**, 4103, 1996.

Ziaka, Z.D., R.G. Minet, and T.T. Tsotsis, *AIChE J.*, **39**, 526, 1993.

Chapter 7

REACTIVE CRYSTALLIZATION

Vaibhav V. Kelkar, Ketan D. Samant, and Ka M. Ng

7.1 Introduction

Reactive crystallization involves simultaneous reaction and solid–liquid phase separation. An example is the liquid-phase air oxidation of *para*-xylene to produce terephthalic acid and another is the absorption of ammonia in aqueous sulfuric acid to form ammonium sulfate. There are three broad issues in the design of reactive crystallization processes. The first involves the chemical plant scale. Here, we consider the design of a complete reactive crystallization flowsheet for systems with multiple reactions and multiple phases, based on an analysis of a reactive solid–liquid equilibrium (SLE) phase diagram. By choosing the proper reaction conditions and flowsheet configurations, the desired product(s) can be manufactured with temperature and pressure swings, while suppressing the formation of undesirable byproducts. Relevant process paths can be shown on the phase diagram as an aid for process conceptualization. The second broad issue is related to the reactive crystallizer itself. Here, reaction kinetics, heat and mass transfer effects, and crystallization kinetics are considered. Identification of the dominant mechanism(s) in the crystallizer provides an understanding of how these factors impact the equilibrium-based plant-scale design and a rational overall plan for operating the reactive crystallizer. Next, various mixing scales (micromixing, mesomixing, and macromixing) are considered. The aim is to provide a more detailed plan for operating the crystallizer. These are now discussed in turn below.

7.2   Solid–Liquid Phase Diagrams with Reactions

7.2.1   Generation of Phase Diagrams

Traditionally, thermodynamic data used for design of reactive crystallization are primarily limited to solubility. While this may be adequate for a
system with a single reaction and a single precipitating solute, the use of
phase diagrams is essential for multiple-reaction, multiple-solid systems. It
provides a more complete view for identifying the solid(s) that can be
obtained under the specified operating conditions. A solid–liquid phase
diagram with reactions can be generated using the following equations.
Since each reaction satisfies reaction equilibrium, we have

$$K_R = \exp\left(\frac{-\Delta G_R}{\Re T}\right) \tag{7.1}$$

where K_R is the reaction equilibrium constant, ΔG_R is the Gibbs free energy
of reaction R, $\Re$ is the gas constant, and T is the temperature. The equilibrium constant is related to the mole fractions of the species by

$$K_R = \prod_{i=1}^{c} (x_{i,eq}\gamma_i)^{\nu_{i,R}} \tag{7.2}$$

where $x_{i,eq}$, γ_i, and $\nu_{i,R}$ are the equilibrium composition, activity coefficient,
and stoichiometric coefficient for component i in reaction R, respectively.
When the solid phases are completely immiscible, the saturation composition can be calculated using the solubility equation

$$x_i^{sat} = \frac{1}{\gamma_i} \exp\left(\frac{\Delta H_{m,i}^0}{\Re}\left(\frac{1}{T_{m,i}} - \frac{1}{T}\right)\right) \tag{7.3}$$

where ΔH_m^0 is the heat of melting, and T_m the melting temperature for
component i. Notice that for ideal systems, the solubility of component i
is a function of only temperature. The following algorithm was proposed by
Berry and Ng (1997) for calculating condensed, nonionic solid–liquid phase
diagrams with a single liquid phase:

1. Specify values for T and P.
2. Choose k component(s) to be saturated.
3. Initialize γ_i, $i = 1, 2, \ldots, c$.
4. Calculate K_R, $R = 1, 2, \ldots, r$ by equation 7.1.
5. Choose values less than x_j^{sat} for the remaining $(c - r - k)$ compositions.
6. Calculate the saturation composition of the k solids by equation 7.3.
7. Calculate remaining r compositions with equation 7.2.

8. Note that the sum of the mole fractions must add up to unity.
9. Calculate γ_i, $i = 1, 2, \ldots, c$.
10. Repeat steps 6–9 until the latest estimate of mole fractions equal their previous values.
11. If the solution indicates that the specified solids cannot coexist or one or more of the $c-k$ components are saturated, there are three options. One, make a new guess for the values of the unsaturated components and then go to step 5. Two, choose a different set of k solids and go to step 3. Three, specify a new temperature and go to step 2.
12. Accept point.
13. Repeat entire algorithm with a new temperature, a different set of k components or with new values of x_j so that saturation boundaries of the k components are clearly designated.

7.2.2 Example Phase Diagrams

Figure 7.1(a) shows a system composed of components A, B, and P, which undergoes the following reaction:

$$A + B \leftrightarrow P \downarrow \tag{7.4}$$

where P has limited solubility in the solution. Line 15 represents the solubility limit of P. A composition within triangle $15P$ represents a mixture with the presence of solid P. The region below line 15 ($15BA$) consists of only liquid mixtures. Thus, point 1 is the single saturation point of P in A whereas point 5 is the single saturation point of P in B. The curve $A6B$ represents the reaction equilibrium curve at the given temperature. The

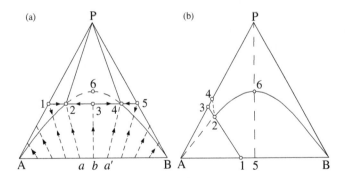

Figure 7.1: Heterogeneous phase diagram for (a) $A + B \leftrightarrow P \downarrow$ (after Berry and Ng, 1997) and (b) $A_{(s)} + B \leftrightarrow P$.

reaction equilibrium curve intersects the solubility curve at points 2 and 4. These are the fixed points of the system where both reaction and phase equilibrium are satisfied. Because of stoichiometry, the composition of a mixture changes in a predetermined direction as reaction proceeds. These directional stoichiometric lines are shown as dashed lines in the phase diagram. For example, point b is an equimolar feed of A and B, and the stoichiometric line points towards vertex P. This stoichiometric line intersects the solubility curve at point 3, which is a saddle point. As the system attempts to attain reaction equilibrium represented by point 6, theoretically, A and B would react to extinction to form solid P. Points a and a' would lead to fixed points 2 and 4. For a mixture with an initial composition between a and b, the reaction path follows the stoichiometric line until the intersection with the solubility line. At that point, the reaction path makes a left turn and ends at the fixed point 2. Similarly, for a mixture with an initial composition between a' and b, one ends up at fixed point 4. For an initial composition within region $A12a$ ($a'45B$), it ends up at the reaction equilibrium curve $A2$ ($B4$), without forming solid P. For an initial composition within triangle $13P$, the system tends towards point 2. In contrast, for a mixture within region $35P$, the system tends towards point 4.

Figure 7.1(b) shows a system composed of components A, B, and P, which undergoes the following reaction:

$$A_{(s)} + B \leftrightarrow P \tag{7.5}$$

where A has limited solubility in the solution. It represents a system where a solid reacts with a liquid. Line 13 represents the solubility of A at the given temperature. Curve $A2B$ is the reaction equilibrium curve. Point 2 is the fixed point of the system where the solubility curve and the reaction equilibrium curve intersect. Line 24 is the segment of the stoichiometric line which passes through point 2 and lies above the reaction equilibrium curve. Any initial composition in the region $124PB$ will result in a liquid mixture with a composition represented by a point on the reaction equilibrium curve. For example, any initial composition along the stoichiometric line $5P$ will end up at point 6. A mixture with a composition within the triangle 234 will proceed along the applicable stoichiometric line, then the solubility line 32, and end up at point 2. For an initial composition within the triangle $A13$, the system tends to point 2.

7.2.3 Representation of High-Dimensional Phase Diagrams

These calculations can be extended to systems with four or more components. The major challenge lies in selecting the proper representation for such high-dimensional systems in process synthesis. Cuts and projections can be used for this purpose. A cut is a section of the phase diagram obtained by fixing one of the coordinate variables for the phase diagram. Actually, Figures 7.1(a) and (b) are isothermal cuts in that the temperature is fixed at a specified value. Note that successive cuts can be made to a phase diagram. For example, in Figure 7.1(a), if we further specify the mole fraction of *P* to correspond to that of the solubility line 15, the phase diagram becomes a line with two fixed points, 2 and 4.

A projection is a phase diagram obtained by renormalizing or transforming the coordinate variables. Figure 7.2 shows an isothermal quaternary phase diagram and its Jänecke projection. The apex I represents the pure solvent. There are three solubility surfaces, one for each of the components *A*, *B*, and *P*. A single liquid phase exists above the solubility surfaces. There are two double-saturation points, *AB*, *AP*, and *BP*, and a triple-saturation point *ABP*. At each double-saturation point, as well as the double-saturation line joining this point to the triple-saturation point, two solutes would precipitate out. At the triple-saturation point, all three solutes precipitate out. These features can be projected onto the base of the tetrahedron to obtain the Jänecke projection. Physically, the projection is obtained by viewing the solubility surfaces through the apex I. Mathematically, the compositions are renormalized by eliminating the mole or mass fraction of solvent I.

A projection can also be obtained using transformed coordinates. Ung and Doherty (1995) provided a general transform to plot multicomponent vapor–liquid phase diagrams of reacting systems. The same transform was used by Slaughter and Doherty (1995) to plot binary solid–liquid phase diagrams for systems which undergo solid-phase reactions to form compounds and by Berry and Ng (1997) for multicomponent systems which

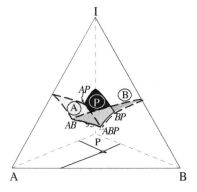

Figure 7.2: Isothermal tetrahedral phase diagram for a simple-eutectic, three-solute system. (After Dye and Ng, 1995.)

undergo liquid-phase reactions to form solids. Physically, the coordinate system is chosen in such a way that all the compositions shown in a transformed phase diagram are those on a reaction surface. Using this transformed coordinate system reduces the number of independent coordinates by the number of reactions without loss of any information from a phase diagram plotted in mole fraction coordinates. Phase diagrams of multicomponent systems with three or fewer degrees of freedom can easily be plotted regardless of the number of components or the number of reactions.

The transform may be derived in the following manner. Consider a system of r independent reactions among c reacting components. The chemical reactions can be represented in the following way:

$$\nu_{1R}A_1 + \nu_{2R}A_2 + \cdots + \nu_{cR}A_c = 0 \qquad (R = 1, 2, \ldots, r) \qquad (7.6)$$

where A_i are the reacting species, and ν_{iR} is the stoichiometric coefficient of component i in reaction R. By convention $\nu_{iR} < 0$ if component i is a reactant in reaction R, $\nu_{iR} > 0$ if component i is a product, and $\nu_{iR} = 0$ if component i does not participate in reaction R. Using vector–matrix formalism, the following vectors are defined:

$$\mathbf{v}_i^T = (\nu_{i1}, \nu_{i2}, \ldots, \nu_{ir}) \qquad (7.7)$$

$$\mathbf{v}_{TOT}^T = (\nu_{TOT1}, \nu_{TOT2}, \ldots, \nu_{TOTr}) \qquad (7.8)$$

where $\mathbf{v}_i^T$ is the row vector of stoichiometric coefficients of component i in each reaction and $\mathbf{v}_{TOT}^T$ is the row vector of the sum of the stoichiometric coefficients for each reaction. At any instant, the mole fraction of component i, x_i, can be expressed as

$$x_i = \frac{x_i^0 + \mathbf{v}_i^T \boldsymbol{\xi}}{1 + \mathbf{v}_{TOT}^T \boldsymbol{\xi}} \qquad (i = 1, \ldots, c) \qquad (7.9)$$

where x_i^0 is the initial mole fraction of component i and $\boldsymbol{\xi}$ is the column vector of the r dimensionless extents of reaction

$$\boldsymbol{\xi} = (\xi_1, \xi_2, \ldots, \xi_r)^T \qquad (7.10)$$

The r extents of reaction are eliminated from these c equations by choosing a subsystem of r equations from equation 7.9. The r components chosen to eliminate the r extents of reaction are called the reference components, thus,

$$\mathbf{x}_{Ref} = \frac{\mathbf{x}_{Ref}^0 + \mathbf{V}\boldsymbol{\xi}}{1 + \mathbf{v}_{TOT}^T \boldsymbol{\xi}} \qquad (7.11)$$

where $\mathbf{x}_{Ref}$ is the column vector composed of the mole fractions of the r reference components, $\mathbf{x}_{Ref}^0$ is the column vector of the initial mole fractions of the r reference components, and $\mathbf{V}$ is the square matrix of stoichiometric coefficients for the r reference components in the r reactions, or

$$\mathbf{x}_{Ref} = \left(x_{(c-r+1)}, x_{(c-r+2)}, \dots, x_c\right)^T \tag{7.12}$$

$$\mathbf{x}_{Ref}^0 = \left(x_{(c-r+1)}^0, x_{(c-r+2)}^0, \dots, x_c^0\right)^T \tag{7.13}$$

$$\mathbf{V} = \begin{pmatrix} \nu_{(c-r+1)1} & \cdots & \nu_{(c-r+1)r} \\ \cdots & \nu_{ir} & \cdots \\ \nu_{c1} & \cdots & \nu_{cr} \end{pmatrix} \tag{7.14}$$

By substituting ξ from equation 7.11 into equation 7.9, one obtains the following transformed set of coordinates:

$$X_i = \frac{x_i^0 - \mathbf{v}_i^T \mathbf{V}^{-1} \mathbf{x}_{Ref}^0}{1 - \mathbf{v}_{TOT}^T \mathbf{V}^{-1} \mathbf{x}_{Ref}^0} = \frac{x_i - \mathbf{v}_i^T \mathbf{V}^{-1} \mathbf{x}_{Ref}}{1 - \mathbf{v}_{TOT}^T \mathbf{V}^{-1} \mathbf{x}_{Ref}} \quad (i = 1, \dots, c-r) \tag{7.15}$$

The $(c-r)$ transforms possess two convenient properties. First, they sum to unity. Second, the transformed variables are reaction invariant, or

$$X_i(0) = X_i(\xi) \quad \forall \xi \quad (i = 1, \dots, c-r) \tag{7.16}$$

This means that the transformed coordinates describe the system as if no reactions were occurring at all.

Figure 7.3 shows a mole fraction ternary phase diagram in which $A + B \leftrightarrow P$. An isothermal cut at 300 K shows that components A and P can precipitate out as solids at that temperature while B remains in solution. The dashed line is the reaction equilibrium curve which intersects the solubility line for component A at a single point. When the temperature is lowered to 270 K, component B can also precipitate out. The reaction equi-

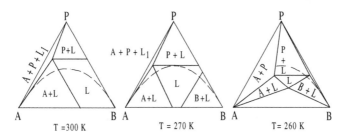

Figure 7.3: Isothermal phase diagrams with a liquid-phase reaction $A + B \leftrightarrow P$. (After Berry and Ng, 1997.)

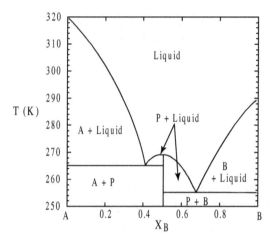

Figure 7.4: Polythermal phase diagram with transformed coordinates for a system with an intermediate melting compound. (After Berry and Ng, 1997.)

librium curve intersects the solubility lines for A and B, and touches the solubility line for P. If we reduce the temperature slightly below 270 K, the reaction equilibrium curve is expected to intersect the P solubility line at two points. At 260 K, the reaction equilibrium curve no longer intersects the solubility line for A. Notice that the reaction equilibrium curve is not extended within the solid–solid–liquid region because only the liquid-phase reaction is considered.

Such a series of isothermal cuts can be conveniently viewed in transformed coordinates. Figure 7.4 shows that at 300 K, A and a liquid are in reaction and solid–liquid equilibrium corresponding to the single intersection point in Figure 7.4 at the same temperature. The transformed variable X_B is given by $(x_B + x_P)/(1 + x_P)$. At 270 K, a liquid can coexist with A, B, or P. At 260 K, only B and P can coexist with a liquid, again under reaction and solid–liquid equilibrium conditions.

7.2.4   Process Synthesis Based on Phase Diagrams

The advantage of using the transformed phase diagrams becomes even more evident for systems with multiple phases and multiple reactions. Consider the following reactions in the presence of a volatile inert I:

$$A + B \leftrightarrow P \downarrow \qquad\qquad (7.17)$$

$$A + D_{(g)} \leftrightarrow E \downarrow \qquad\qquad (7.18)$$

Both P and E have limited solubility in the solution. Component P is the desired product. Component E is an unwanted byproduct. Gas I can be used to purge D to a lower vapor concentration and thus reduce the concentration of D in solution. The number of degrees of freedom where P equilibrates with solution and vapor is $6(c) - 3(p) - 2(r) = 1$. The number of transformed variables is 4. The temperature is fixed and pressure is 1 atm. Figure 7.5 shows the system in terms of X_D and X_I. Components P and E are chosen to be the reference component. Compositions within region PL_13 split into solution and crystals of P. This analysis clearly shows the conditions under which only P crystallizes.

Irrespective of the chosen representation, these phase diagrams can be used to synthesize processes to recover pure components from reactive mixtures. For example, Figure 7.6 shows the phase diagram for acetic acid, quinaldine, and isoquinoline and a flowsheet to separate the components. Stream numbers in the flowsheet and points on the phase diagram correspond to one another. The feed is combined with the effluent from C2, stream (4), to make stream (1). The stream is cooled to T_1. Crystals of isoquinoline precipitate from solution and are then filtered from the process. The effluent, stream (2) is combined with the distillate to make stream (3). The stream is fed to C2 which operates at T_2. Crystals of the quinaldine–acetic acid adduct are filtered, remelted, and sent to a distillation column. Assuming that the quinaldine does not decompose, the bottoms are composed of quinaldine at a specified purity. The distillate rich in acetic acid is recycled.

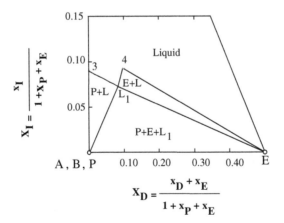

Figure 7.5: Phase diagram for a vapor–liquid–solid system with two reactions. (After Berry and Ng, 1997.)

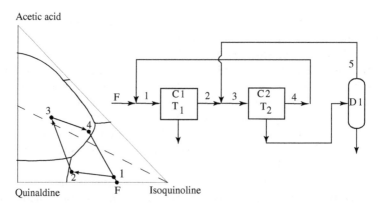

Figure 7.6:   Process for separating quinaldine and isoquinoline. (After Berry and Ng, 1997.)

7.3 Crystallization Kinetics

Apart from the reaction kinetics and the mass transfer processes observed in single or multiphase reactors, reactive crystallizers involve two additional distinct rate processes, which together constitute the kinetics of crystallization. New nuclei are formed in solution, which then grow to form crystals. The nucleation and crystal growth steps are the least understood of all the steps occurring in a reactive crystallizer, though they are the dominating factor in deciding the crystallizer performance, such as the amount and the crystal size distribution of the crystallizing solute. The driving force for both processes is the departure of the chemical potential of the solute in the supersaturated solution, μ, from its equilibrium value, μ^*. This driving force is usually approximated as a difference in the concentration, C, in the supersaturated solution, and C^*, in the saturated solution. The supersaturation is commonly expressed either as the absolute supersaturation, $\Delta C = C - C^*$, the relative supersaturation, $\sigma = \Delta C / C^*$, or the supersaturation ratio, $S = C/C^*$. In each case, the concentration itself may be expressed in one of several different units (e.g., molarity, mole fraction, molality, mass fraction, etc.), leading to different values of ΔC, σ, and S for the same system (Mullin and Söhnel, 1977). The kinetics of nucleation and crystal growth are briefly discussed below. For a short review that traces the developments in crystallization kinetics from the earliest stages, see Hulbert (1984).

7.3.1 Kinetics of Nucleation

Nucleation facilitates the phase transition of a solute from the liquid to the solid phase in a supersaturated solution. Nuclei may be born due to several mechanisms, each of which results in a distinct nucleation rate expression. The reader is referred to the excellent text by Nyvlt et al. (1985) for a detailed discussion. The important mechanisms are briefly discussed below.

7.3.1.1 Primary Nucleation

This can be further classified as homogeneous or heterogeneous primary nucleation. Homogeneous nucleation results in the formation of nuclei directly from the solution, and in the absence of any solid phase to facilitate the nucleation process. Heterogeneous nucleation is induced by a solid surface other than that of the crystallizing solute itself. This may be the walls of the crystallizer, the heat transfer surface, the agitator, etc.

The classical theory of nucleation of liquid drops from the vapor phase (e.g., Burton, 1977) may be adapted to the homogeneous primary nucleation of a solid phase from a supersaturated solution. This theory assumes that solute molecules in supersaturated solutions gradually aggregate to form clusters. With increasing size of these clusters, their overall free energy increases, until at a certain critical size, it is maximum. This critical cluster size, where the clusters have an equal probability of growth or disintegration, is given by (Mersmann, 1984).

$$\frac{x^*}{2} = \frac{2\gamma^s M_w}{\Re T \rho_s \ln(1 + \Delta C/C^*)} \tag{7.19}$$

For clusters above size x^*, the free energy decreases. Thus, nuclei with diameters $x < x^*$ are unstable, while those above the critical size are stable, and are capable of further growth into crystals.

For typical values of surface tension, γ^s, and the supersaturation, ΔC, observed in crystallizing systems, Mersmann (1984) estimated the size of the critical nucleus to lie between 1 and 1000 nm. He thus classified the nuclei and crystals according to the following criteria:

$x < 1\,\mu m$: metastable nuclei
$1\,\mu m < x < 50\,\mu m$: "subsieve crystals" or microcrystals
$x > 50\,\mu m$: sieve-size crystals

Reactive crystallization or precipitation processes typically operate at a high relative supersaturation, $\Delta C/C^*$, leading to a low size of the critical nuclei, and hence a larger number of stable nuclei. This may explain the higher nucleation rate observed in precipitation systems, compared to that observed in crystallization by cooling and evaporation. The nucleation rate for the mechanism of primary nucleation is the rate at which the clusters overcome the free energy barrier, and is given in Table 7.1. For a heterogeneous nucleation mechanism the rate expression is of a similar form, but with different multiplying constants, because of the favorable energetics of forming a critical nucleus on a foreign solid surface, rather than in solution.

7.3.1.2 Secondary Nucleation

When new nuclei are formed in a supersaturated solution due to the presence of the solid phase of the crystallizing solute itself, the mechanism is termed secondary nucleation. Secondary nucleation may be further classified into several types (Söhnel and Garside, 1992). Tiny crystallites detached from the seed crystals may act as crystallization centers (apparent secondary nucleation), new nuclei may be formed as a result of crystal–solution interaction (true secondary nucleation), or, crystal–crystal, crystal–impeller, and crystal–wall collisions may give rise to new nuclei (contact nucleation). The

Table 7.1:   Crystallization Kinetic Models.

Nucleation kinetic models

Homogeneous primary nucleation (classical theory)	$\alpha_1 \exp\left[-\dfrac{\alpha_2}{T^3(\ln S)^2}\right]$	Nyvlt et al. (1985)
Secondary nucleation	$k_n' \Delta C^n M_t^j \omega^b$	Garside and Davey (1980)

Crystal growth kinetic models

Diffusion-controlled growth	$\dfrac{k_a}{3k_v}\dfrac{k_S \Delta C}{\rho_s}$	Mersmann (1984)
Mononuclear mechanism	$\alpha_1 \sigma^{1/2} \exp\left[-\dfrac{\alpha_2}{T^2\sigma}\right]$	
Polynuclear mechanism	$\alpha_1 \sigma^{3/2} \exp\left[-\dfrac{\alpha_2}{T^2\sigma}\right]$	Nyvlt et al. (1985)
"Birth and spread" model	$\alpha_1 \sigma^{5/6} \exp\left[-\dfrac{\alpha_2}{T^2\sigma}\right]$	
Screw dislocation mechanism (BCF theory)	$\alpha_1 T^2 \exp\left[-\dfrac{\alpha_2}{T}\right]\sigma^2 \tanh\left[\dfrac{\alpha_3}{T\sigma}\right]$	

kinetics of secondary nucleation are usually correlated in terms of the super-saturation, ΔC, the magma density, M_t, and the agitation speed, ω, giving the nucleation rate expression shown in Table 7.1.

For systems of readily soluble substances, contact nucleation appears to be the most important secondary nucleation mechanism (Garside and Davey, 1980). For systems of sparingly soluble solutes, such as in precipitation processes, the role of secondary nucleation is insignificant in comparison with that of primary nucleation, and secondary nucleation can usually be ignored (Söhnel and Garside, 1992; Mersmann and Kind, 1988). This is because the size of crystals formed during precipitation is too small for secondary nucleation mechanisms to play an important role. Over limited ranges of supersaturation, the dependence of nucleation rate on the super-saturation can be modeled using a power law expression of the form

$$J_n = k_n(C - C^*)^n \qquad (7.20)$$

Though the parameters k_n and n have no physical significance, expressions of the above form are commonly used because of their convenience. Table 7.2 shows the values of k_n and n for selected industrial reactive crystallization systems.

7.3.2   Kinetics of Crystal Growth

The different faces of a crystal grow at different velocities. Hence, crystal growth is usually defined in terms of an overall linear growth rate G, which is given as the rate of increase of the radius of a sphere having a volume equal to the average volume of one crystal. Numerous theories for crystal growth can be found in the literature; some consider crystal growth from a purely thermodynamic point of view, while others deal with the actual kinetics of crystal growth (Nyvlt et al., 1985). Some of the important theories are discussed below.

The growth process is often separated into two steps—the mass transport of the dissolved solute from the bulk of the solution to the crystal face, and the subsequent surface integration of the solute molecules into the crystal lattice. If growth is completely controlled by the diffusion step, the overall growth rate is simply proportional to the liquid–solid mass transfer coefficient, and the concentration difference between the bulk and the crystal face (see Table 7.1).

However, for small supersaturations, small relative velocities between the solution and the crystals, and low temperatures, the surface integration step usually dominates the overall growth rate (Mersmann, 1984). Surface integration of the solute molecule may be achieved by several mechanisms, some of which are schematically shown in Figure 7.7.

Table 7.2: Kinetic Parameters for Selected Industrial Reactive Crystallization Systems.

		Growth Kinetics		
Compound	C^* (mol m^{-3})	k_g (m s^{-1})(mol m^{-3})$^{-g}$	g	Reference
BaSO$_4$	0.01	3×10^{-8}	2	Mersmann and Kind (1988)
CaC$_2$O$_4$	0.052	2.07×10^{-8}	2	Nielsen (1984)
CaCO$_3$	0.065	1.09×10^{-8}	2	Nielsen and Toft (1984)
BaCO$_3$	0.25	1.04×10^{-9}	2	Mersmann and Kind (1988)
BaC$_2$O$_4$	0.49	1.16×10^{-8}	2	Nielsen (1984)
CaHPO$_4 \cdot$ 2H$_2$O	1.76	2.4×10^{-10}	2	Christoffersen and Christoffersen (1988)
MgC$_2$O$_4 \cdot$ 2H$_2$O	4.71	9.91×10^{-14}	2	Nielsen (1984)

		Nucleation Kinetics		
Compound	C^* (mol m^{-3})	k_n (s m^{-3})(mol m^{-3})$^{-n}$	n	Reference
CaCO$_3$	0.065	107	4.2	Wachi and Jones (1991)
HO $\cdot$ C$_6$H$_4 \cdot$ COOH (salicylic acid)	10.4	2×10^{-3}	6.85	Franck et al. (1988)

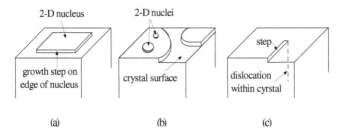

Figure 7.7: Crystal growth due to (a) mononuclear mechanism, (b) "birth and spread" mechanism, and (c) screw dislocation mechanism.

Figure 7.7(a) shows a growth due to a mononuclear mechanism, where a two-dimensional nucleus is formed on the surface of the crystal, and grows instantaneously to form a complete new crystal layer before another nucleus is formed. A polynuclear mechanism refers to a growth due to the formation of several two-dimensional nuclei on the same crystal surface, until the whole surface is covered, while the nuclei themselves do not grow. Figure 7.7(b) shows a growth according to the "birth and spread" model (Gilmer and Bennema, 1972) which allows formation as well as growth of more than one nucleus on the crystal surface. A new crystal layer is formed when the growing nuclei merge into each other. Figure 7.7(c) shows growth by a screw dislocation mechanism (Burton et al., 1951; Bennema, 1967). Growth may then occur for example, due to the propagation of the dislocation around its axis. The growth rate expressions resulting from these have been summarized in Table 7.1.

In spite of the considerable theoretical advances made in predicting the relationship between crystal growth and parameters such as the temperature and supersaturation, it is still not possible to accurately determine the growth mechanism prevailing in a particular system under a certain set of conditions. Values of a number of parameters occurring in the kinetic equations cannot be determined with sufficient accuracy. Hence, for engineering purposes the growth rate kinetics are often empirically modeled as a power law expression such as

$$G = k_g(C - C^*)^g \tag{7.21}$$

where k_g is the growth rate constant, and g is the kinetic order of growth. The parameters k_g and g are obtained by fitting experimental data. Table 7.2 shows the values of the growth kinetic parameters for selected industrial reactive crystallization systems. Söhnel and Garside (1992) report additional data from experimental precipitation studies of a large number of compounds. Garside and Shah (1980) have reviewed the applicability of empirical expressions for representing nucleation and growth kinetics in MSMPR crystallizers.

7.4 The Population Balance

Crystallization is a particulate process and any description of a crystallizing system is incomplete without a population balance, which characterizes the particulate material by the distribution of size of its particles. The particle size distribution (PSD) can be the major determining factor in the ultimate use of a particulate material; hence, it is one of the most important design objectives in a crystallization process. Randolf and Larson (1962) first formalized rational techniques for the prediction of PSD, based on the population balance in crystallizers. The PSD data are commonly presented in several different ways, such as the cumulative totals (or cumulative fractions), or the density, of a measured quantity such as the crystal number, area, or mass, plotted against the particle size. A few commonly used distributions are shown in Figure 7.8 (Randolf and Larson, 1988).

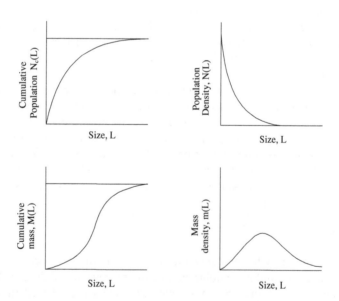

Figure 7.8: Common forms for representing a PSD: the cumulative population and mass distributions, and the population and mass density distributions.

The exact shape of the distribution depends on the crystallizer system, and is obtained by solving the population balance equation. The complete population balance for a size range ∂L, for a crystallizer under transient conditions can be given as

$$\frac{\partial N}{\partial t} + \frac{\partial (GN)}{\partial L} + D(L) - B(L) + N\frac{\partial (\ln V)}{\partial t} + \frac{F}{V}(N - N_i) = 0 \qquad (7.22)$$

This includes the change in population density of the crystals due to growth, death, birth, a change in volume of the suspension, and flows being fed and discharged from the crystallizer. Several techniques for the solution of population balance equations are available in the literature (e.g., Ramkrishna, 1971; Kumar and Ramkrishna, 1996). For a steady-state, constant-volume MSMPR crystallizer with no crystals in the feed, and a growth rate independent of the crystal size, the above population balance reduces to

$$G\tau\frac{\partial N}{\partial L} + N = 0 \qquad (7.23)$$

where $\tau = V/F$. This leads to a simple expression for the number density, N

$$N = N_0 \exp(-L/(G\tau)) \qquad (7.24)$$

where N_0 is the population density of the nuclei. Equation 7.24 is the basis for determination of growth kinetics from experimental studies in a MSMPR crystallizer (Jancic, 1982).

Other distributions such as the length, area, and mass distribution are obtained as higher moments of the number distribution. The mass or weight fraction distribution is particularly useful and has the following form for an MSMPR crystallizer:

$$w(L) = \frac{L^3 \exp(-L/(G\tau))}{6(G\tau)^4} \qquad (7.25)$$

The maximum of the weight fraction distribution may be defined as the dominant particle size of the distribution, and is given by

$$L_d = 3G\tau \qquad (7.26)$$

The PSD changes when the growth rate is size dependent (Canning and Randolf, 1967; Abegg et al., 1968), or when growth dispersion exists, i.e., the growth rate is different for different particles of the same size (Larson et al., 1985). It is also affected by several other factors such as seeding, batch or continuous operation, mixed or classified product removal, etc. It is now also possible to incorporate the PSD models for units such as crushers, dissolvers, and filters, along with that for the crystallizer, and track the PSD of a solids stream from one unit to the other in a complete plant

(Hill and Ng, 1997). For a review of the advances made in modeling and control of the PSD in crystallizers, see Randolf (1984) and Randolf and Larson (1988).

7.5   Generic Model for a Reactive Crystallizer

Reactive crystallization is a complex phenomenon involving several processes which take place simultaneously. Reaction in solution results in supersaturation of the dissolved product, which leads to the formation of crystal nuclei. Mass transfer of the dissolved solute to existing crystals and incorporation of the solute into the crystal lattice leads to crystal growth. These mechanisms or steps proceed sequentially, and the overall rate is determined by the slowest step. This scenario can be further complicated by a gas–liquid reaction where gas–liquid mass transfer represents an additional step (Wachi and Jones, 1991). In some reactive crystallization processes, one of the reactants may be in solid form (e.g., salting-out crystallization with a solid salting-out agent). Here, the dissolution of the solid reactant represents another step. All the individual steps are linked together through the supersaturation, ΔC, which is generated by reaction, dissipated by mass transfer, and consumed by nucleation and growth. Figure 7.9, based on the one proposed by Garside (1985), shows the feed-

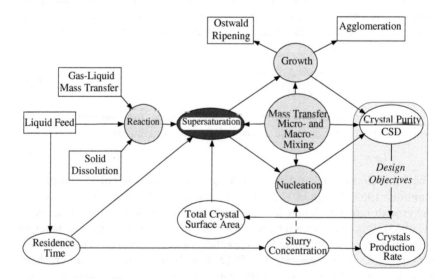

Figure 7.9:   Interactions in a reactive crystallization system.

forward and the feedback nature of the interactions present in a reactive crystallization system. The shaded circles represent the most important steps taking place in a reactive crystallizer, while rectangles denote secondary processes. Ovals are used to denote process variables or crystal attributes. The arrows indicate the sequence of steps, as well as the linkages between the steps, and the process variables and crystal attributes. The dashed arrow between *slurry concentration* and *nucleation* indicates that secondary nucleation plays only a secondary role in reactive crystallization processes.

In recent years, there has been a concerted effort to represent crystallization within the general framework of chemical reaction engineering (Garside and Tavare, 1984). Such an approach involves the development of mass and energy balances for the crystallization system, accounting for any or all of the individual steps mentioned above. Phenomenological models are used for the nucleation and crystal growth processes, rather than seeking a rigorous description of the actual kinetic mechanisms. Although the MSMPR crystallizer continues to be a convenient benchmark for crystallization studies, the use of other models to characterize the nonideal mixing in crystallizers has been explored. These include the tanks-in-series model (Abegg and Balakrishnan, 1971), the dispersed plug flow model (Rivera and Randolf, 1978), and the double draw-off crystallizer model (White and Randolf, 1989). Batch crystallizers have also been analyzed (Tavare et al., 1980; Marchal et al., 1988).

Several investigators have studied the interplay among the individual mechanisms in a reactive crystallization system, and their effect on the resulting CSD. Matsuoka and Garside (1993) discussed the transport of solute from the bulk solution to the crystal–solution interface and its subsequent incorporation into the crystal lattice. Franck et al. (1988) modeled salicylic acid precipitation, accounting for diffusion- or kinetic-limited crystal growth, primary and secondary nucleation, and growth by agglomeration of smaller crystals. Tavare and Garside (1990) investigated the effect of feed addition rate and Ostwald ripening on the crystal characteristics in a semibatch reactive crystallizer with no mass transfer limitations. Wachi and Jones (1991) analyzed a gas–liquid heterogeneous precipitation system, by combining the film theory of gas–liquid mass transfer with the population balance equations for the precipitating solute. The role of heat transfer at the liquid–crystal interface on crystal growth has not been extensively studied. The heat of crystallization released during crystal growth changes the temperature at the liquid–crystal interface, affecting the solubility, as well as the rate constant of the surface integration process. This effect may be strong when the solute concentration is high, and the heat of crystallization is large. Garside and Tavare (1981) and Matsuoka and Garside (1993) discussed this effect in terms of a nonisothermal effectiveness factor for crystal growth. Garside and Tavare (1981) concluded that heat transfer at the liquid–crystal interface is relatively unimportant for crystallization from solution, while Matsuoka et al. (1972) found that it can be significant during crystal growth from melts. All the above studies, however, focus only on a few steps in reactive crystallization. A generic model which accounts for all possible steps is missing. A framework for such a generic model is presented below, for an isothermal MSMPR crystallizer (Kelkar and Ng, 1999).

Consider an MSMPR crystallizer being fed with a liquid feed stream containing reactant species A and B at a volumetric flow rate F. A and B react in the liquid phase in the presence of an inert solvent I. The reaction is reversible in general, even if the forward reaction is much faster than the reverse reaction. An amount of product P in excess of the solubility limit precipitates out of the solution as a solid product:

$$A_{(l)} + B_{(l)} \leftrightarrow 2P \downarrow \qquad (7.27)$$

The reaction rate can generally be represented as a function of the species concentrations,

$$\text{Rate} = \Phi(C) \qquad (7.28)$$

The material balances for the reacting species can be set up as:

Balance for species A,

$$F(C^f_A - C_A) - V\Phi = 0 \qquad (7.29)$$

Balance for species B,

$$F(C^f_B - C_B) - V\Phi = 0 \qquad (7.30)$$

Balance for species P,
- liquid phase:

$$F(C^f_P - C_P) + V\Phi - Vk_S a_T(C_P - C^s_P) - V(k_v \rho_m L_0^3)J_n = 0 \qquad (7.31)$$

- liquid–crystal interface:

$$Vk_S a_T(C_P - C^s_P) - V(a_T/2)\rho_m G = 0 \qquad (7.32)$$

Here, C_P is the concentration of the dissolved solute in the bulk of the liquid, C^s_P is the concentration of the solute at the liquid–crystal interface, and C^*_P is the solubility. The nucleation rate, J_n, and linear crystal growth rate, G, have been transformed into molar units by using appropriate multiplication factors. The generic model also includes a population balance (equation 7.22), which can be simplified according to the assumptions made in the generic model.

The above crystallizer model is generic and more details can be added. For example, for a gas–liquid reaction or a system with a solid reactant, the gas–liquid mass transfer and the solid dissolution steps can be modeled using overall mass transfer coefficients. Depending on the mechanism of nucleation and growth, suitable phenomenological expressions could be chosen to represent J_n and G. Effects like Ostwald ripening, agglomeration, and attrition can be included in the birth and death terms of the population balance.

Examination of equations 7.29 to 7.32 reveals four distinct rate steps: reaction between A and B, the generation of P nuclei in the liquid phase, the mass transfer of dissolved P to the growing P crystals, and the surface integration of the solute P into the crystal lattice (i.e., the crystal growth step). Each of the above steps is characterized by a characteristic time. One of the steps often controls the overall rate of the reactive crystallization process although the driving forces adjust themselves so that the rate of each individual step is equal to the overall rate at steady state. This controlling step can be identified by carrying out a sensitivity analysis of the generic model. (For example, see Kelkar and Ng (1998) for a similar sensitivity analysis for gas–liquid and gas–liquid–solid catalytic reactions.) Although such a sensitivity analysis is beyond the scope of this chapter, a further insight into the model can be obtained by making the generic model (equations 7.29–7.32) dimensionless. The dimensionless numbers that result are the ratios of the characteristic times of the individual steps, and determine the relative rates of two consecutive steps. For example, for a reaction which is first order overall, and for nucleation and growth kinetics which can be represented by conventional power law expressions, the resulting dimensionless numbers are shown in Table 7.3.

By examining an appropriate product of dimensionless numbers, it is also possible to identify the slower of any two nonconsecutive steps. For example, the ratio N_{Gr}/Da_M compares the growth rate to the reaction rate; the product $Da_R N_{Nu}$ compares the nucleation rate with the reactor throughput. A generic model as above represents a coherent approach for treating reactive crystallization systems involving any number of components and the different primary (reaction, nucleation, mass transfer, growth) as well as secondary (Ostwald ripening, agglomeration, attrition) phenomena occurring in the crystallizer.

7.6 Mixing in Reactive Crystallizers

The generic model presented earlier accounts for the steps of reaction, mass transfer from the bulk to the crystal face, nucleation, and crystal growth

Table 7.3: Dimensionless Numbers in the Generic Model.

Dimensionless Number	Definition	Expression
Reaction Damköhler number	$\dfrac{\text{Reaction rate}}{\text{Reactor throughput}}$	$\text{Da}_R = k_1 \tau$
Mass transfer Damköhler number	$\dfrac{\text{Reaction rate}}{\text{Mass transfer rate (bulk to crystal)}}$	$\text{Da}_M = \dfrac{k_1}{k_S a_T}$
Nucleation number	$\dfrac{\text{Nuclei generation rate}}{\text{Reaction rate}}$	$N_{Nu} = \dfrac{(K_v \rho_m L_0^3) k_n C_P^{*n-1}}{k_1}$
Growth number	$\dfrac{\text{Surface integration rate}}{\text{Mass transfer rate (bulk to crystal)}}$	$N_{Gr} = \dfrac{(a_T/2) \rho_m k_g C_P^{*g-1}}{k_S a_T}$

Reaction rate $\Phi = k_1(C_A - (C_P/K_R))$.
Nucleation rate $J_n = k_n(C_P - C_P^*)^n$.
Growth rate $G = k_g(C_P^s - C_P^*)^g$.

phenomena occurring within a reactive crystallizer. However, this model assumes that the bulk liquid phase in the reactive crystallizer is perfectly mixed. In other words, the level of supersaturation is assumed to be the same at all locations within the crystallizer. More often than not, this assumption is not very accurate. Mixing controls instantaneous chemical reaction and hence the generation of supersaturation of the crystallizing product(s). Local variations in the level of supersaturation can affect the nucleation and growth processes in different manners, thus influencing various crystal attributes such as mean particle size, particle size distribution, and particle morphology. It is almost impossible to predict the influence of mixing on crystal attributes intuitively. It is necessary to understand the interplay between mixing, reaction, nucleation, and crystal growth within a reactive crystallizer. Therefore, we conclude this chapter with a brief review of experimental and modeling studies that investigate the effect of mixing on reactive crystallization.

The first part of this review deals with experimental studies. It presents a summary of various experimental studies and observations made which are of importance in the design and performance evaluation of industrial crystallizers. In the second part, we discuss the modeling methods that treat turbulent mixing, reaction, nucleation, and growth processes simultaneously. Here, we discuss models for mixing, the modeling scheme for a reactive crystallization, and the length and time scales of mixing. In the third part, we identify various operating regimes and present a few examples of the effects of various operating conditions on the performance of a reactive crystallizer in these regimes.

7.6.1 Experimental Studies

Effect of mixing on the performance of reactive crystallization processes has been extensively studied experimentally in the recent past. The emphasis of these experimental studies is on demonstrating the effect of mixing on various crystal attributes and on observing the effects of changes in various operating conditions on these crystal attributes. A summary of these experimental studies is provided in Table 7.4. This summary presents information on the source, the process considered, the mode of operation, the operating conditions varied, and the crystal attributes studied. It should be noted that this summary is not exhaustive and represents only a cross-section of the information available in the literature. The interested reader is referred to David and Marcant (1994) and Aslund and Rasmuson (1992) for additional listings.

It is interesting to note that in these experimental studies, many apparently contradictory observations have been reported. For instance,

Table 7.4: Summary of Experimental Studies on the Effect of Mixing on Reactive Crystallization Processes.

Reference	Process Studied	Mode of Operation	Operating Conditions Varied	Crystal Attributes Studied
Pohorecki and Baldyga (1983)	Barium sulfate	Batch (single feed)	Agitator rpm, reactant concentration	Mean particle size
Kuboi et al. (1986)	Barium sulfate	Semibatch (single and double feed)	Feed location	Mean particle size
Franck et al. (1988)	Salicylic acid	Continuous	Residence time	Mean particle size
Pohorecki and Baldyga (1988)	Barium sulfate	Continuous (single feed) Continuous (double feed)	Agitator rpm Agitator rpm	Mean particle size Mean particle size
Tosun (1988)	Barium sulfate	Semibatch (double feed)	Agitator rpm, feed location	Mean particle size
Baldyga et al. (1990)	Barium sulfate	Semibatch (double feed)	Agitator rpm	Mean particle size
Fitchett and Tarbell (1990)	Barium sulfate	Continuous (double feed)	Agitator rpm	Mean particle size

Mydlarz et al. (1991)	Zinc oxalate	Continuous (single feed)	Agitator rpm, residence time	Mean particle size
Aslund and Rasmuson (1992)	Benzoic acid	Semibatch (single feed)	Agitator rpm, feed location, feed rate, reactant concentration, impeller type	Mean particle size, particle morphology
Baldyga et al. (1995)	Barium sulfate	Semibatch (double feed)	Agitator rpm, feed location, feed rate, reactant concentration	Mean particle size, particle size distribution, particle morphology
Franke and Mersmann (1995)	Calcium carbonate	Batch (double feed)	Mean power input, batch time	Mean particle size
	Calcium carbonate Calcium sulfate	Continuous (double feed)	Mean power input, feed location and concentration, residence time	Mean particle size, particle morphology
Chen et al. (1996)	Barium sulfate	Semibatch (single feed)	Agitator rpm, feed location, feed rate	Mean particle size, particle size distribution

Pohorecki and Baldyga (1983) observed a decrease in mean particle size of barium sulfate crystals with an increase in impeller speed for batch precipitation. The same authors (1988) studied the precipitation of barium sulfate in continuous stirred tank reactors using premixed and unpremixed feed. In both cases, mean crystal size increased with increase in impeller speed. Tosun (1988) and Baldyga et al. (1990) observed a minimum on the crystal size vs. impeller speed curve. Chen et al. (1996) reported similar observations for feed locations near the impeller. For feed locations away from the impeller, they observed that the mean particle size goes through a maximum and a minimum as impeller speed increases.

These and other reported observations are very important for the design of industrial crystallizers, but do not have very simple explanations. Such trends are observed due to the interaction of mixing on various scales and the influence of these scales on reaction, nucleation, and growth phenomena. To understand and explain these results, we need to be able to model turbulent mixing in conjunction with molecular-level phenomena such as reaction, nucleation, and growth. In the following section, we discuss techniques for the same.

7.6.2   Modeling of Turbulent Mixing in Reactive Crystallizers

In liquid-phase turbulent-flow systems, widely different length scales are involved in the mixing process that has to bring about molecular level processes. The macroscale is of the order of the vessel itself. Mixing on this scale is the turbulent dispersion of feed from inlet throughout the vessel. On the other hand, micromixing occurs at much smaller scales— the Kolomogoroff microscale. Molecular phenomena such as diffusion, and accompanying chemical reactions, nucleation, and crystal growth take place at this scale.

In any model of turbulent mixing in a reactive crystallizer all scales must be completely described. Ideally, this would involve solving the Reynolds averaged Navier–Stokes equations with appropriate closure approximations, source and sink terms that account for reaction, nucleation, and growth, and boundary conditions in conjunction with the population balance equations described earlier. This is an extremely tedious task and is impossible to perform with the current state-of-the-art. Hence, phenomenological models are commonly used to describe the microscale phenomena, and the macroscales are calculated by using either computational fluid dynamics (CFD) models or experimentally constructed flow models.

For a phenomenological model of micromixing to be useful, it should be based as closely as possible on the physical understanding of the actual process. It should also be able to account for the interactions between micro- and macroscales. Several models, such as environment (E) models, the interaction by exchange with mean (IEM) model, and the engulfment deformation diffusion (EDD) model, are available in the literature (Villermaux and Falk, 1994). The EDD model (and the simplified E model) proposed and developed by Bourne's research group (Baldyga and Bourne, 1984, 1988, 1989, 1992) is one of the most widely used models and is based on the understanding of key physical processes contributing to mixing on molecular scale. Therefore, in the following sections we briefly describe the mechanisms, and the scales of mixing associated with this model.

It is interesting to note here that several other models have been used to study the effect of mixing on reactive crystallization processes. For example, Becker and Larson (1969), and Garside and Tavare (1985) discuss the performance of a reactive crystallizer at the limits of micromixing (perfect mixing and complete segregation on molecular scale). Cell balance models which assume the crystallizer to be divided into a large number of perfectly mixed cells which are connected according to the macroscale flow pattern have also been used. While these studies offer useful insights, they cannot fully explain the effect of various operating conditions on the crystal attributes. A summary of other modeling studies is presented by Tavare (1986).

7.6.2.1 EDD Model Mechanisms

Engulfment, deformation, and diffusion are the principal mechanisms responsible for effecting mixing on molecular level (Baldyga and Bourne, 1984). Turbulent vorticity mixes the fluid by rolling up and therefore creating multiple layers or slabs of fresh fluid and fluid from its environment. This process of incorporation of fluid from the environment into deforming vortices is called engulfment. The scale of segregation in these vortices is reduced by deformation in the turbulent flow field. Molecular diffusion (assumed unidirectional) accompanied by molecular scale processes (reaction, nucleation, and growth) takes place within these deforming vortices.

7.6.2.2 Scales of Mixing

Micromixing. Micromixing takes place by the EDD mechanism as described above. For Sc < 4000 (as is the case with many reactive crystallization systems), engulfment is the controlling mechanism for micromixing. The rate of engulfment can be expressed as

$$E = 0.05776(\varepsilon/v)^{0.5} \tag{7.33}$$

This rate takes different values at different locations within a reactive crystallizer as the turbulence is inhomogeneous. The time scale for micromixing can be written using the average rate of turbulent energy dissipation as

$$t_M = E_{av}^{-1} = \frac{1}{0.05776} \left(\frac{\nu}{\varepsilon_{av}}\right)^{0.5} \tag{7.34}$$

Mesomixing. Mesomixing refers to the coarse-scale turbulent exchange between fresh feed and its surroundings. For fast chemical reactions, the reactions are localized near the feed point. Thus, the plume of fresh feed is of a coarse scale relative to the micromixing scale and of fine scale relative to the scale of the vessel. This scale of the plume is called the mesoscale. Near the feed inlet, fresh feed spreads out by turbulent dispersion. Mesomixing brought about by turbulent dispersion and related models for interaction between meso and micromixing are described by Baldyga and Bourne (1992). The characteristic time scale of mesomixing was defined (assuming point source) as

$$t_D = \frac{F}{\bar{u}D_t} \tag{7.35}$$

Macromixing. As mentioned earlier, macromixing refers to the spreading or circulation of the fresh feed throughout the reactor. In a reactive crystallizer, the values of velocities and rates of turbulent energy dissipation differ locally. Thus macroscale motion not only determines the compositions of the environment with which the reacting vortices interact but also determines their positions and local turbulence properties at these locations as they move through the vessel. Macroscale flows can be calculated either by using commercial CFD codes (Bakker and Van Den Akker, 1994) or by using experimentally developed flow models (Bourne and Yu, 1994). The characteristic time scale for macromixing is the circulation time (t_C).

7.6.2.3 Modeling Scheme

The source and sink terms in EDD model equations involve terms for generation and dissipation of supersaturation through reaction, nucleation, and crystal growth. The rate expressions used for these processes are as discussed in the preceding section. In addition, we need to solve population balance equations to keep track of mean particle size and particle size distribution. Thus the complete scheme of modeling mixing in reactive crystallizers is as

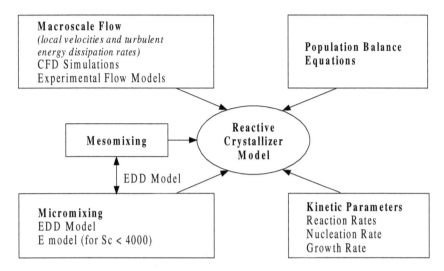

Figure 7.10: Modeling scheme for effect of mixing on reactive crystallization processes.

shown schematically in Figure 7.10. Details of model equations can be found in papers such as Baldyga et al. (1995).

7.6.3 Operating Regimes

From a design standpoint, it is extremely important to be able to classify the modeling results and experimental observations into various operating regimes. The dimensionless numbers and time scales useful in characterizing the operating regimes of a reactive crystallizer are listed in Table 7.5. Let us now discuss these operating regimes and the trends in crystal attributes that can be expected in these regimes.

7.6.3.1 *Perfectly Mixed Regime*

In a reactive crystallizer, mixing at various scales occurs simultaneously with reaction, nucleation, and growth processes. However, if micromixing is much faster than the other molecular level processes that it brings about, the contents of the reactive crystallizer can be assumed to be perfectly mixed. The above condition translates to

$$M_i \ll 1 \qquad (i = 1, \ldots, 4) \tag{7.36}$$

Table 7.5: Time Scales and Dimensionless Numbers Used to Characterize Operating Regimes.

Time scales

t_M = Characteristic time of micromixing
t_D = Characteristic time of mesomixing
t_C = Characteristic time of macromixing (circulation time)
t_R = Characteristic time of reactor operation
$t_R = t_F$ = Feed addition time (semibatch operation)
$t_R = \tau$ = Average residence time (continuous operation)

Dimensionless numbers

$$M_1 = \mathrm{Da}_R \frac{t_M}{t_R} = \frac{\text{Characteristic time of micromixing}}{\text{Characteristic time of reaction}}$$

$$M_2 = \frac{\mathrm{Da}_R}{\mathrm{Da}_M} \frac{t_M}{t_R} = \frac{\text{Characteristic time of micromixing}}{\text{Characteristic time of mass transfer (bulk to crystal)}}$$

$$M_3 = \frac{\mathrm{Da}_R}{N_{Nu}} \frac{t_M}{t_R} = \frac{\text{Characteristic time of micromixing}}{\text{Characteristic time of nucleation}}$$

$$M_4 = \frac{\mathrm{Da}_R}{\mathrm{Da}_M} N_{Gr} \frac{t_M}{t_R} = \frac{\text{Characteristic time of micromixing}}{\text{Characteristic time of growth (surface integration)}}$$

$$Q = \frac{t_D}{t_M} = \frac{\text{Characteristic time of mesomixing}}{\text{Characteristic time of micromixing}}$$

Note: The Damköhler numbers for reaction and mass transfer and the nucleation and growth numbers are as defined earlier in Table 7.3.

If all of these conditions are satisfied, mixing essentially precedes reaction, nucleation, and crystal growth. Hence, the generic model developed earlier can be used without any loss of information. Also, in the perfectly mixed regime, operating conditions, such as impeller type, agitator rpm, feed location, and feed addition time in the case of semibatch operation, have no effect on crystal attributes.

7.6.3.2 *Mixing-Controlled Regime*

If any one or more of the conditions given above are not satisfied, the molecular level processes are affected by mixing. In this mixing-controlled regime, one has to use the modeling scheme discussed in section 7.6.2. The mixing-controlled regime can be further subdivided on the basis of the dominant length scale of mixing as discussed below:

- *Micromixing-controlled regime.* Mixing on the microscale controls when

$$Q = t_D/t_M \ll 1 \tag{7.37}$$

  In this regime, turbulent dispersion of the feed from the inlet is not important and can be completely neglected.

- *Mesomixing-controlled regime.* This regime is characterized by

$$Q = t_D/t_M \geq 1 \tag{7.38}$$

  In this regime, the time scale of mesomixing is comparable to or greater than the time scale of micromixing. The turbulent dispersion of feed from the inlet controls the process of mixing. Self-engulfment (engulfment of fresh fluid with itself) becomes increasingly important as the value of Q increases.

- *Macromixing-controlled regime.*

$$Q = t_D/t_M \gg 1 \quad \text{or} \tag{7.39a}$$

$$t_R/t_C \ll 1 \tag{7.39b}$$

  Here, the turbulent dispersion of feed spreads the entire vessel. The modeling scheme discussed earlier is less trustworthy in this regime as the macroscale flows are calculated separately, independent of the microscale phenomena.

7.6.4   Effect of Operating Conditions on Crystal Attributes

In this section, let us consider some examples of how various operating conditions influence the crystal attributes in various operating regimes.

Example 1: Effect of power input per unit volume (or agitator rpm). Let us consider continuous operation in the micromixing-controlled regime of a system for which the rate of nucleation is much higher than the rates of reaction and growth. Precipitation of calcium carbonate (Franke and Mersmann, 1995) is an example of such a system. As the power input per unit volume (or equivalently agitator rpm) increases, the supersaturation is dissipated more rapidly due to increased rate of engulfment of fresh feed with its environment. This leads to a substantial decrease in the rate of nucleation and the number of nuclei formed. Hence the mean particle size increases as the power input per unit volume increases. If the power input is made sufficiently high, mixing essentially precedes all molecular processes (t_M decreases and equations 7.36 are satisfied) and the MSMPR limit, as shown by the dashed line in Figure 7.11(a), for particle size is reached.

Example 2: Effect of agitator rpm and feed location. Let us consider a system for which growth rate is higher than the nucleation rate at very high supersaturations but decreases very rapidly with decrease in supersaturation. Let us also consider the reactive crystallizer operating in semibatch mode in the micromixing-controlled regime. Here we will evaluate the effect of varying the agitator rpm on the mean particle size for two different feed locations; in the impeller plane, and near the free surface.

In the impeller plane, the rate of turbulent energy dissipation and hence the rate of engulfment are high enough to substantially reduce the supersaturations at all agitator rpm values. Hence, at all agitator rpm values, the rate of nucleation is higher than the rate of growth in the impeller plane. As the agitator rpm is increased from a low value,

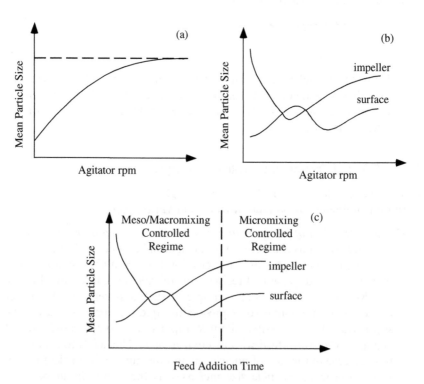

Figure 7.11:   Effect of operating conditions on mean particle size.

the rate of nucleation decreases. However, the time for completion of the reaction increases, thus allowing more time for nucleation. The net effect is an increase in the number of nuclei formed and a decrease in mean particle size. As the agitator rpm is increased further, the increase in the time of nucleation is not sufficient to compensate for the decrease in the rate of nucleation. The net effect is a decrease in the number of nuclei formed and an increase in average particle size. This is shown schematically in Figure 7.11(b) and explains the experimental and modeling results of Pohorecki and Baldyga (1983), Tosun (1988), and Chen et al. (1996) for barium sulfate precipitation.

At locations near the free surface, the local dissipation of supersaturation is not so effective, since the local rates of engulfment are substantially low. Hence at low rpm values, the growth rate is higher than the nucleation rate. As the rpm is increased from a lower value, both these rates decrease. However, initially the growth rate is still higher than the rate of nucleation, thus giving larger crystals. As the rpm is increased further, the nucleation rate becomes greater than the growth rate and the mean particle size goes through a minimum as explained, see Figure 7.11(b), and as observed by Chen et al. (1996).

Similar observations were made by Aslund and Rasmuson (1992) for semibatch precipitation of benzoic acid in a single-feed reactive crystallizer.

Example 3: Effect of feed addition time (or feed rate). Let us consider the same system as in Example 2. As the feed addition time for semibatch operation decreases, we move from the micromixing-controlled regime into the mesomixing- and macromixing-controlled regimes. In these regimes there is increased self-engulfment and the dissipation of supersaturation is not as effective as in the micromixing-controlled regime. Hence decrease in the feed addition time has the same qualitative effect on the mean crystal size as the decrease in agitator rpm. Thus, we expect to see the trends similar to the previous example at different feed locations, see Figure 7.11(c). This is indeed as observed by Chen et al. (1996).

For continuous operation, the mean particle size is expected to vary in a similar fashion as the average residence time is varied.

7.6.5 Choice of Operating Regimes

One may operate the reactive crystallizer in different operating regimes by properly selecting various reactor attributes and operating conditions such as agitator type and rpm, feed concentration (for second- and higher-order reactions), feed location, feed addition rate (for semibatch operation), and average residence time (for continuous operation). These attributes are related directly or indirectly to the various operating regimes through the time scales and dimensionless numbers as shown in Table 7.5 (also see equations for time scales). Thus an ideal choice of reactor attributes and operating conditions can be made easily if the desired operating regime is known.

As discussed in the previous subsection, the modeling results give us the performance of a given system in various operating regimes. However, from a design standpoint, it is essential to be able to get a good idea of which operating regime is suitable for any given system, without doing extensive modeling. A conceptual framework which can treat this aspect is thus necessary for successful design of industrial reactive crystallizers.

7.7 Conclusions

In this review, a multiscale approach is proposed for the design of reactive crystallization processes. At the plant scale, we design the process from which the desired product(s) will be produced. This is accomplished by examining the reactive solid–liquid phase diagrams. At the reactive crystallizer scale, the impact of transport limitations and kinetics on the equilibrium-based design is considered. Dimensionless numbers characterizing the relative importance of reaction kinetics, heat and mass transfer, and crystallization kinetics are used to identify the dominant step(s). The equipment design and operation specifications including impeller type, feed addition time, feed addition location, etc., are further tightened by incorporating the various mixing scales in the design. These three scales should not be viewed in isolation. Instead, they should be considered concurrently, albeit with the emphasis shifting from the large scale to the small scale as the project progresses. We believe that this approach will lead to an optimal design in a timely fashion as demanded in this competitive environment of the global chemical processing industries.

Symbols

a_T	Surface area of growing crystals per unit suspension volume, $m^2\,m^{-3}$
$A, B, \ldots$	Reactant species
c	Number of components
$C1, C2$	Crystallizers
C_i	Concentration in the liquid bulk, $mol\,m^{-3}$

C_P^*	Solubility of P in the solution, mol m^{-3}
C_P^s	Concentration of P at the liquid–crystal interface, mol m^{-3}
$B(L), D(L)$	Birth and death terms in the population balance (equation 7.22)
$D1$	Distillation column
D_t	Turbulent diffusivity, m^2 s^{-1}
Da_M	Damköhler number for mass transfer of solute from bulk liquid to the crystal face
Da_R	Reaction Damköhler number
E	Rate of engulfment, s^{-1}
E_{av}	Average rate of engulfment, s^{-1}
f	Degrees of freedom
F	Feed stream; volumetric flow rate, m^3 s^{-1}
g	Order of the crystal growth process (equation 7.21)
G	Linear crystal growth rate, m s^{-1}
ΔG_R	Gibbs free energy of reaction R, J mol^{-1}
ΔH_m^0	Heat of melting, J mol^{-1}
J_n	Nucleation rate, number per m^3 s
k_1	First-order reaction rate constant, s^{-1}
k_g	Crystal growth rate constant (equation 7.21), (m s^{-1})/(mol m^{-3})^g
k_n	Nucleation rate constant (equation 7.20), number per m^3 s (mol m^{-3})^n
k_n'	Secondary nucleation rate constant
k_S	Solid–liquid mass transfer coefficient, m s^{-1}
k_v	Volumetric shape factor
K_R	Reaction equilibrium constant
L	Crystal size, mm
L_d	Dominant crystal size, mm
L_0	Size of nuclei, mm
$m(L)$	Mass density distribution of crystals, kg m^{-4}
$M(L)$	Cumulative mass distribution of crystals, kg m^{-3}
M_t	Magma density, kg m^{-3}
M_w	Molecular weight of solute, mol kg^{-1}
n	Order of nucleation rate (equation 7.20)
$N, N(L)$	Population density distribution of crystals, number per m^4
N_i	Population density distribution of crystals in the feed, number per m^4
$N_c(L)$	Cumulative number distribution of crystals, number per m^3
N_{Gr}	Growth number
N_{Nu}	Nucleation number
p	Number of phases
P	Pressure; product of reaction; solute

Q	Ratio of characteristic times of mesomixing and micromixing
$\mathfrak{R}$	Gas constant, J mol^{-1} K^{-1}
S	Supersaturation ratio, C/C^*
Sc	Schmidt number
t	Time, s
t_M, t_D, t_C	Characteristic time for micro-, meso-, and macromixing, respectively, s
t_F	Feed addition time, s
t_R	Characteristic time for reactor operation, s
T	Temperature, K
$\bar{u}$	Local velocity at feed addition point, m s^{-1}
V	Suspension volume, m^3
$\mathbf{V}$	Square matrix of stoichiometric coefficients for reference species
$w(L)$	Weight fraction distribution of crystals, m^{-1}
x_i	Liquid-phase mole fraction of component i
x^*	Critical cluster size
$\mathbf{x}_{Ref}$	Column vector of reference component mole fractions (equation 7.11)
X	Transformed coordinate

Greek

$\alpha_1, \alpha_2, \ldots$	Constants in the crystallization kinetic expressions
ΔC	Supersaturation, mol m^{-3}
ε	Rate of turbulent energy dissipation per unit mass, m^2 s^{-3}
ε_{av}	Average rate of turbulent energy dissipation per unit mass, m^2 s^{-3}
Φ	Molar reaction rate, mol s^{-1}
γ	Activity coefficient
γ^s	Surface tension at the liquid–crystal interface, J m^{-2}
μ	Chemical potential of solute in solution, J mol^{-1}
ν	Kinematic viscosity, m^2 s^{-1}
ν_i	Stoichiometric coefficient of i in r
$\mathbf{v}_i^T$	Row vector of stoichiometric coefficients for each reaction
$\mathbf{v}_{TOT}^T$	Row vector of sum of stoichiometric coefficients for each reaction
ρ_s, ρ_m	Mass and molar density of crystals, mol m^{-3}
σ	Relative supersaturation, $\Delta C/C^*$
τ	Residence time of liquid in the crystallizer, s
ω	Stirring speed, rpm
ξ	Extent of reaction
$\boldsymbol{\xi}$	Column vector of the R dimensionless extents of reaction

Subscripts and superscripts

*	At saturation conditions
eq	Equilibrium
i	Pertaining to species *i*
f, 0	Pertaining to inlet (feed) or initial conditions
m	Melt
Ref	Reference component
s, *sat*	Pertaining to the liquid–crystal interface

References

Abegg, C.F. and N.J. Balakrishnan, "The Tanks in Series Concept as a Model for Imperfectly Mixed Crystallizers," in *Crystallization from Solution: Factors Influencing Size Distribution*, M.A. Larson, Ed., *AIChE Symp. Ser.*, **110**, No. 67, Amer. Inst. Chem. Eng., New York, 1971, p. 88.

Abegg, C.F., J.D. Stevens, and M.A. Larson, *AIChE J.*, **14**, 118, 1968.

Aslund, B.L. and A.C. Rasmuson, *AIChE J.*, **38**, 328, 1992.

Bakker, R.A. and H.E.A. Van Den Akker, *Trans. Inst. Chem. Eng.*, **72**, Part A, 733, 1994.

Baldyga, J. and J.R. Bourne, *Chem. Eng. Commun.*, **28**, 259, 1984.

Baldyga, J. and J.R. Bourne, *Chem. Eng. Sci.*, **43**, 107, 1988.

Baldyga, J. and J.R. Bourne, *Chem. Eng. J.*, **42**, 83, 1989.

Baldyga, J. and J.R. Bourne, *Chem. Eng. Sci.*, **47**, 1839, 1992.

Baldyga, J., R. Pohorecki, W. Podgorska, and B. Marcant, "Micromixing Effects in Semibatch Precipitation," in *Proceedings of the 11th Symposium on Industrial Crystallization*, A. Mersmann, Ed., Garmisch-Partenkirchen, RFG, 175, 1990.

Baldyga, J., W. Podgorska, and R. Pohorecki, *Chem. Eng. Sci.*, **50**, 1281, 1995.

Becker, G.W. and M.A. Larson, *Chem. Eng. Prog., Symp. Ser.*, **65** (95), 14, 1969.

Bennema, P., *J. Crystal Growth*, **1**, 278, 1967.

Berry, D.A. and K.M. Ng, *AIChE J.*, **43**, 1737, 1997.

Bourne, J.R. and S. Yu, *Ind. Eng. Chem. Res.*, **33**, 41, 1994.

Burton, J.J., *Statistical Mechanics*, B.J. Berne, Ed., Plenum, New York, 1977.

Burton, W.K., N. Carberra, and F.C. Frank, *Philos. Trans. R. Soc. London*, **243**, 299, 1951.

Canning, T.F. and A.D. Randolf, *AIChE J.*, **13**, 5, 1967.

Chen, J., C. Zheng, and G. Chen, *Chem. Eng. Sci.*, **51**, 1957, 1996.

Christoffersen, J. and M.R. Christoffersen, *J. Crystal Growth*, **87**, 41, 1988.

David, R. and B. Marcant, *AIChE J.*, **40**, 424, 1994.

Dye, S. and K.M. Ng, *AIChE J.*, **41**, 2427, 1995.

Fitchett, D.E. and J.M. Tarbell, *AIChE J.*, **36**, 511, 1990.

Franck, R., R. David, J. Villermaux, and J.P. Klein, *Chem. Eng. Sci.*, **43**, 69, 1988.

Franke, J. and A. Mersmann, *Chem. Eng. Sci.*, **50**, 1737, 1995.

Garside, J., *Chem. Eng. Sci.*, **40**, 3, 1985.

Garside, J. and R.J. Davey, *Chem. Eng. Commun.*, **4**, 393, 1980.

Garside, J. and M.B. Shah, *Ind. Eng. Chem. Process Des. Dev.*, **19**, 509, 1980.

Garside, J. and N.S. Tavare, *Chem. Eng. Sci.*, **36**, 863, 1981.

Garside, J. and N.S. Tavare, "Crystallization as Chemical Reaction Engineering," in *Inst. Chem. Eng. Symp. Ser.*, No. 87 (ISCRE 8), 767, 1984.

Garside, J. and N.S. Tavare, *Chem. Eng. Sci.*, **40**, 1485, 1985.

Gilmer, G.H. and P. Bennema, *J. Crystal Growth*, **13-14**, 148, 1972.

Hill, P.J. and K.M. Ng, *AIChE J.*, **43**, 715, 1997.

Hulbert, H.M., "Perspectives on Crystallization in Chemical Process Technology," in *Advances in Crystallization from Solutions*, G.R. Youngquist, Ed., AIChE Symp. Ser., **80**, No. 240, Amer. Inst. Chem. Eng., New York, 1984, p. 1.

Jancic, S.J., "Derivation of Design Oriented Crystallization Kinetics from Population Density Data: A Standardized Procedure," in *Industrial Crystallization '81*, S.J. Jancic and E.I. De Jong, Eds., North-Holland, Amsterdam, 1982, p. 11.

Kelkar, V.V. and K.M. Ng, *AIChE J.*, **44**, 1563, 1998.

Kelkar, V.V. and K.M. Ng, *AIChE J.*, **45**, 69, 1999.

Kuboi, R., M. Haranda, J.M. Winterbottom, A.J.S. Anderson, and A.W. Nienow, "Mixing Effects in Double Jet and Single Jet Precipitation," in *Proceedings of the 3rd World Congress of Chemical Engineering*, Tokyo, Japan, Paper 8g-302, 1986.

Kumar, S. and D. Ramkrishna, *Chem. Eng. Sci.*, **51**, 1311, 1996.

Larson, M.A., E.T. White, K.A. Ramanarayanan, and K.A. Berglund, *AIChE J.*, **31**, 90, 1985.

Marchal, P., R. David, J.P. Klein, and J. Villermaux, *Chem. Eng. Sci.*, **43**, 59, 1988.

Matsuoka, M. and J. Garside, *J. Crystal Growth*, **129**, 385, 1993.

Matsuoka, M., S. Fujita, and T. Hayakawa, *Heat Transfer - Jpn. Res.*, **1**, 58, 1972.

Mersmann, A., *Int. Chem. Eng.*, **24**, 401, 1984.

Mersmann, A. and M. Kind, *Chem. Eng. Technol.*, **11**, 264, 1988.

Mullin, J.W. and O. Söhnel, *Chem. Eng. Sci.*, **32**, 683, 1977.

Mydlarz, J., J. Reber, D. Briedis, and A.J. Voigt, *AIChE Symposium Series*, **87**, 158, 1991.

Nielsen, A.E., *J. Crystal Growth*, **67**, 289, 1984.

Nielsen, A.E. and J.M. Toft, *J. Crystal Growth*, **67**, 278, 1984

Nyvlt, J., O. Söhnel, M. Matuchova, and M. Broul, *The Kinetics of Industrial Crystallization*, Elsevier, Amsterdam, 1985.

Pohorecki, R. and J. Baldyga, *Chem. Eng. Sci.*, **38**, 79, 1983.

Pohorecki, R. and J. Baldyga, *Chem. Eng. Sci.*, **43**, 1949, 1988.

Ramkrishna, D., *Chem. Eng. Sci.*, **26**, 1134, 1971.

Randolf, A.D., "Advances in Crystallizer Modeling and CSD Control," in *Advances in Crystallization from Solutions*, G.R. Youngquist, Ed., AIChE Symp. Ser., **80**, No. 240, Amer. Inst. Chem. Eng., New York, 1984, p. 14.

Randolf, A.D. and M.A. Larson, *AIChE J.*, **8**, 639, 1962.

Randolf, A.D. and M.A. Larson, *Theory of Particulate Processes*, 2nd Edition, Academic, London, 1988.

Rivera, T. and A.D. Randolf, *Ind. Eng. Chem. Process Des. Dev.*, **17**, 182, 1978.

Slaughter, D.W. and M.F. Doherty, *Chem. Eng. Sci.*, **50**, 1679, 1995.

Söhnel, O. and J. Garside, *Precipitation: Basic Principles and Industrial Application*, Butterworth-Heinemann, Oxford, 1992.

Tavare, N.S., *AIChE J.*, **32**, 705, 1986.

Tavare, N.S. and J. Garside, *Trans. Inst. Chem. Eng.*, **68**, 115, 1990.

Tavare, N.S., J. Garside, and M.R. Chivate, *Ind. Eng. Chem. Process Des. Dev.*, **19**, 653, 1980.

Tosun, G., "An Experimental Study of the Effect of Mixing on the Particle Size Distribution in Barium Sulfate Precipitation Reaction," in *Proceedings of the 6th European Conference on Mixing*, Pavia, Italy, 161, 1988.

Ung, S. and M.F. Doherty, *Chem. Eng. Sci.*, **50**, 23, 1995.

Villermaux, J. and L. Falk, *Chem. Eng. Sci.*, **49**, 5127, 1994.

Wachi, S. and A.G. Jones, *Chem. Eng. Sci.*, **46**, 1027, 1991.

White, E.T. and A.D. Randolf, *Ind. Eng. Chem. Res.*, **28**, 276, 1989.

INDEX

reaction processes (*cont.*)
 esterification, 3, 12, 20–21, 64–65, 84, 132, 137, 156, 181–183
 etherification, 3, 12, 22, 24, 42, 45, 47, 84
 fat splitting, 58, 62
 hydroformylation, 78–81, 171
 hydrogenation, 24, 42, 120, 124, 133, 152, 156, 164–165, 171, 173, 175
 hydrolysis, 12, 58–61, 63–64, 68, 86, 118, 138, 140, 142, 152, 186
 isomerization of light paraffins, 144
 methane oxidative coupling, 149
 methanol synthesis, 136
 methanolysis, 58
 methyl acetate hydrolysis, 140
 methyl acetate process, 20–21, 37
 methyl acetate synthesis, 138
 nitration, 12, 52, 68–70
 oxidation, 16, 52, 74, 81, 86, 156, 159, 165–168, 170–173, 180, 192, 195, 209
 oximation, 75, 77
 PUREX™ process, 56–57, 87
 saponification, 12, 58, 62–63, 72
 selectivity-limited reactions, 144
 SHOP™ process, 72, 77–78
 steam reforming, 168, 174, 192, 195
 transesterification, 2, 66
reactive separation equipment
 chromatographic reactor, 118
 countercurrent moving-bed chromatographic reactor, 120
 membrane reactor, 5, 15, 155–181, 183–185, 187, 191–192, 194
 rotating cylindrical annulus chromatography reactor, 118
 simulated countercurrent moving-bed chromatographic reactor, 121–122
 trickle-bed reactor, 130

reactive separation processes
 absorption with reaction, 3, 12–13, 16, 93, 95
 adsorption with reaction, 115–116, 131
 CDTech tertiary amyl methyl ether (TAME) process, 23
 Eastman methyl acetate process, 37
 economic analysis, 179
 economic impact, 151
 hydrometallurgical, 51–52, 57, 87
 reaction kinetics, 131
 reactive absorption, 2–3, 29, 48
 reactive azeotrope, 27
 reactive chromatography, 115
 reactive crystallization, 209, 219, 221–223, 226–227, 229, 232–233, 235, 237, 242
 reactive distillation, 2–4, 6, 8, 12, 16, 18–20, 22–25, 27–28, 31, 33, 35–42, 44–45, 47, 183
 reformulated gasoline, 22, 144
 UOP Merox™ process, 74
ricinoleic acid, 63

salicylic acid, 222, 227
soaps, 51, 58, 72, 77
sucrose fatty acid esters, 66
sulfur, 14, 74, 80, 107, 132

tertiary amyl methyl ether (TAME), 23
toluene diisocyanate, 69
tributyl phosphate, 53
triglycerides, 58, 60, 72
triphenylphosphine, 61, 79
 trisulfonated, 79

uranium, 66, 67

vitamin K_1, 63–64

zirconium, 53–54

Milton Keynes UK
Ingram Content Group UK Ltd.
UKHW040108071024
449327UK00019B/901

9 780367 447168